POLITICS AND ECONOMICS OF THE MIDDLE EAST

IRAN: SOCIAL, ECONOMIC AND POLITICAL DEVELOPMENTS

POLITICS AND ECONOMICS OF THE MIDDLE EAST

Additional books in this series can be found on Nova's website under the Series tab.

Additional E-books in this series can be found on Nova's website under the E-books tab.

POLITICS AND ECONOMICS OF THE MIDDLE EAST

IRAN: SOCIAL, ECONOMIC AND POLITICAL DEVELOPMENTS

JOSHUA A. OWENSTEIN
EDITOR

Nova Science Publishers, Inc.
New York

For permission to use material from this book please contact us:
Telephone 631-231-7269; Fax 631-231-8175
Web Site: http://www.novapublishers.com

LIBRARY OF CONGRESS CATALOGING-IN-PUBLICATION DATA

Iran : social, economic, and political developments / editor, Joshua A.
Owenstein.
p. cm.
Includes index.
ISBN 978-1-60741-552-7 (hardcover)
1. Iran--Politics and government--1997- 2. Iran--Economic
conditions--1997- 3. Nuclear energy--Iran. 4. Nuclear arms control--Iran.
5. Ballistic missiles--Iran. I. Owenstein, Joshua A.
JQ1785I73 2009
955.06'1--dc22
201002697

Published by Nova Science Publishers, Inc. ✦New York

CONTENTS

PREFACE

Much of the debate over U.S. policy toward Iran has centered on the nature of the current regime; some believe that Iran, a country of about 70 million people, is a threat to U.S. interests because hardliners in Iran's regime dominate and set a policy direction intended to challenge U.S. influence and allies in the region. This book explores Iran's political history, regime structure, ballistic missile program, economic conditions, stability, and presidential elections.

Chapter 1 - features a profile of Iran.

Chapter 2 - President Obama has said his Administration shares the goals of the previous Administration to contain Iran's strategic capabilities and regional influence, but the Obama Administration formulated approaches to achieve those goals that differ from those of its predecessor—in particular through expanded direct diplomatic engagement with Iran. This effort was put into practice with messages to the Iranian people by President Obama, and through invitations to and contact with Iranian diplomats at multilateral meetings, including those on Iran's nuclear program. The Administration also de-emphasized or discontinued policies that Iranian leaders considered hostile including: ratcheting up international sanctions, efforts to promote democracy in Iran, and openly discussing the potential for U.S. military action.

The Administration strategy on Iran is in some flux because of the allegations of a "stolen election" by the challengers to declared winner of the June 12, 2009 presidential election, incumbent President Mahmoud Ahmadinejad, and the crackdown against protesters who demanded a new vote. President Obama has criticized Iran's use of violence against protesters, and observers say that the Administration might try to take advantage of Iran's internal weakness to obtain a compromise that curbs Iran's nuclear program. If Iran refuses to return to the nuclear bargaining table, the Administration might focus, at an earlier than expected stage, on sanctioning and pressuring Iran. Bills in the 111th Congress, such as H.R. 2194 and S. 908, would tighten U.S. sanctions on Iran by amending the Iran Sanctions Act to penalize sales to Iran of gasoline and some of the measures contained in these bills have begun to advance as a consequence of the election-related violence. Congress also has passed resolutions that express solidarity with the demonstrators and condemn the regime's repression of them.

The Obama Administration has not changed the previous Administration's characterization of Iran as a "profound threat to U.S. national security interests," a perception generated primarily by Iran's nuclear program and its military assistance to armed groups in

Iraq and Afghanistan, to the Palestinian group Hamas, and to Lebanese Hezbollah. The U.S. policy approach was to try to prevent a nuclear breakout by Iran primarily by applying progressive multilateral economic pressure on Iran while also offering it potential cooperation should it suspend its enrichment of uranium. Multilateral efforts to pressure Iran include three U.N. Security Council resolutions (1737, 1747, and 1803) that ban weapons of mass destruction (WMD)-related trade with Iran; freeze the assets of Iran's nuclear entities and personalities; prevent Iran from transferring arms outside Iran; ban international travel by some Iranians; call for inspections of some Iranian sea and airborne cargo shipments; and call for restrictions on dealings with some Iranian banks. Separate U.S. efforts to persuade European governments to curb trade, investment, and credits to Iran; and to convince foreign banks not to do business with Iran, began to weaken Iran's economy, compounding the effect of lower oil prices since mid-2008. To emphasize its approach, the George W. Bush Administration built a substantial naval presence in the Persian Gulf.

Much of the debate over U.S. policy toward Iran has centered on the nature of the current regime; some believe that Iran, a country of about 70 million people, is a threat to U.S. Minterests because hardliners in Iran's regime dominate and set a policy direction intended to challenge U.S. influence and allies in the region. President George W. Bush, in his January 29, 2002, State of the Union message, labeled Iran part of an "axis of evil" along with Iraq and North Korea.

Chapter 3 - On June 12, 2009, following a heated campaign between reformist candidate Mir Hussein Musavi and incumbent President Mahmoud Ahmadinejad, Iranians turned out in record numbers to vote in the presidential election. Shortly after the polls closed, the Interior Minister announced that President Ahmadinejad had been reelected by a 62% margin. The announcement was followed by allegations of vote rigging and election fraud and prompted supporters of leading reformist candidate Mir Hussein Musavi and others to hold public demonstrations in several major cities of a size and intensity unprecedented since the Iranian Revolution of 1979.

Despite a government ban on unauthorized public gatherings, protests have reportedly continued every day since the election. Restrictions on foreign and domestic journalists, reported disruptions of mobile phone networks, limited accessibility of some internet sites, mass arrests, and clashes between civilian protestors and Basij forces have garnered international attention and increased concerns about the Iranian government's apparent disregard for human rights and basic civil liberties.

Regardless of the actual election results, the current stand-off between the government and opponents of the election outcome has caused observers to speculate about how this stalemate will be resolved, and what the outcome might mean for U.S. efforts to resolve the issue of Iran's nuclear weapons program, its support for terrorism, and other national security concerns.

Chapter 4 - Iran has an active interest in developing, acquiring, and deploying a broad range of ballistic missiles, as well as developing a space launch capability. This was spotlighted several times since 2008. In mid-July 2008, Iran launched a number of ballistic missiles during military exercises, reportedly including the medium-range Shahab-3. At the time, a Pentagon spokesman said Iran was "not testing new technologies or capabilities, but rather firing off old equipment in an attempt to intimidate their neighbors and escalate tension in the region." Subsequent analysis of the July 2008 missile launches shows Iran apparently digitally altered images of those launches. Iran announced other missile and space launch tests in August and November 2008. In February 2009, Iran announced it launched a satellite into orbit and "officially achieved a presence in space." This short report seeks to provide an overview of the reported or suspected variety of Iranian ballistic missile programs. Because there remains widespread public divergence over particulars, however, this chapter does not provide specificity to what Iran may or may not have, or is in the process of developing.

Iran's ballistic missile program dates to the late 1970s after the Shah was overthrown and the Islamic Republic of Iran established. The new Iranian government embarked on a ballistic missile program marked by considerable secrecy. Many consider that Iran's ballistic missile development was in full force by the mid-1980s during its protracted war against Iraq, during which Iran reportedly launched more than 600 ballistic missiles.

Today, there is little disagreement among most experts that Iran has acquired some number of ballistic missiles from other countries and has developed other ballistic missiles indigenously or in cooperation with others. Iranian ballistic missile proliferation has been a matter of U.S. and international concern. At the same time, however, there has been considerable public disagreement over precisely what kinds of ballistic missile systems Iran has or is developing itself or in cooperation with others. This is because there is little transparency in Iran's ballistic missile programs, which has led to some degree of a lack of confidence in Iran's public assertions of its activities. Finally, details about Iranian ballistic missile programs remain classified in the United States. Because of the secrecy inherent in the development of weapon systems, especially in less open societies, open-source analyses reflect a wide-range of technical views and assessments.

This chapter provides a brief description of what is publicly discussed regarding Iran's ballistic missile programs[1]; it does not discuss Iranian cruise missiles[2] or rockets.[3] These latter weapons were a source of concern when some reportedly Iranian-made rockets and other missiles may have been used by Hezbollah against Israel in 2006.[4] Charges of Iranian military support to Iraqi insurgents do not include Iranian-made rockets or missiles, however. This chapter first examines Iran's long-range ballistic missile programs because those efforts generally drive the greatest concerns within the United States, especially when coupled with concern over the development of Iran's nuclear capabilities. A brief overview of Iran's medium and short-range ballistic missile programs then follows.

Chapter 5 - The Islamic Republic of Iran, a resource-rich and labor-rich country in the Middle East, is a central focus of U.S. national security policy. The United States asserts that Iran is a state sponsor of terrorism and that Iran's uranium enrichment activities are for the development of nuclear weapons. To the extent that U.S. sanctions and other efforts to change Iranian state policy target aspects of Iran's economy as a means of influence, it is important to evaluate Iran's economic structure, strengths, and vulnerabilities.

Since 2000, Iran has enjoyed broad-based economic growth. However, strong economic performance has been hindered by high levels of inflation and unemployment and low levels of foreign investment. Some contend that President Ahmadinejad's expansionary monetary and fiscal policies have worsened unemployment, inflation, and poverty in Iran. Iran's economic growth is expected to slow in 2009.

Iran has long been subject to U.S. economic sanctions, and more recently, to United Nations sanctions, over its uranium enrichment program and purported support for terror activities. Such sanctions are believed by some analysts to contribute to Iran's growing international trade and financial isolation.

Iran's economy is highly dependent on the production and export of crude oil to finance government spending, and consequently is vulnerable to fluctuations in international oil prices. Although Iran has vast petroleum reserves, the country lacks adequate refining capacity and imports gasoline to meet domestic energy needs. Iran is seeking foreign investment to develop its petroleum sector. While some deals have been finalized, reputational and financial risks may have limited other foreign companies' willingness to finalize deals.

While Iran-U.S. economic relations are limited, the United States has a key interest in Iran's relations with other countries. As some European countries have curbed trade and investment dealings with Iran, other countries, such as China and Russia, have emerged as increasingly important economic partners. Iran also has focused more heavily on regional trade opportunities, such as with the United Arab Emirates.

High oil prices have increased Iran's leverage in dealing with international issues, but the country's dependence on oil and other weak spots in the economy have to come to light by the 2008 international financial crisis, which may portend a slowing down of Iran's economy.

Members of Congress are divided about the proper course of action in respect to Iran. Some advocate a hard line, while others contend that sanctions are ineffective at promoting policy change in Iran and hurt the U.S. economy. In the 110th Congress, several bills were introduced that reflect both perspectives. Policies toward Iran remain a key issue for the 111th Congress.

Chapter 6 - In 2002, the International Atomic Energy Agency (IAEA) began investigating allegations that Iran had conducted clandestine nuclear activities; the agency ultimately reported that some of these activities had violated Tehran's IAEA safeguards agreement. The agency has not stated definitively that Iran has pursued nuclear weapons, but has also not yet been able to conclude that the country's nuclear program is exclusively for peaceful purposes. The IAEA Board of Governors referred the matter to the U.N. Security Council in February 2006. Since then, the council has adopted five resolutions, the most recent of which (Resolution 1835) was adopted in September.

The Security Council has required Iran to cooperate fully with the IAEA's investigation of its nuclear activities, suspend its uranium enrichment program, suspend its construction of a heavy- water reactor and related projects, and ratify the Additional Protocol to its IAEA safeguards agreement. However, a February 2009 report from IAEA Director-General Mohamed ElBaradei to the agency's Board of Governors indicated that Tehran has continued to defy the council's demands by continuing work on its uranium enrichment program and heavy-water reactor program. Iran has signed, but not ratified, its Additional Protocol.

Iran and the IAEA agreed in August 2007 on a work plan to clarify the outstanding questions regarding Tehran's nuclear program. Most of these questions have essentially been resolved, but ElBaradei told the agency's board in June 2008 that the agency still has questions regarding "possible military dimensions to Iran's nuclear programme." According to ElBaradei's report, the IAEA has not been able to make progress on these matters.

This chapter provides a brief overview of Iran's nuclear program and describes the legal basis for the actions taken by the IAEA board and the Security Council.

In: Iran Social, Economic and Political Developments
Editor: Joshua A. Owenstein
ISBN: 978-1-60741-552-7

Chapter 1

COUNTRY PROFILE: IRAN*

Library of Congress

COUNTRY

Formal Name: Islamic Republic of Iran (Jomhuri-ye Eslami-ye Iran).

Short Form: Iran.

Term for Citizen(s): Iranian(s).

Capital: Tehran.

Other Major Cities (in order of population): Mashhad, Esfahan, Tabriz, Karaj, Shiraz, Qom, and Ahvaz.

Independence: In the modern era, Iran always has existed as an independent country.

Public Holidays: The national holiday, Islamic Revolution Day, celebrates the victory of the Islamic Revolution on February 11, 1979. Other official holidays are Dawn (the return of Ayatollah Khomeini from 14 years of foreign exile on February 1, 1979), Noruz (Iranian New Year, March 21), Islamic Republic Day (April 1), Thirteenth Day of New Year (April 2), and several Islamic religious holidays that are reckoned in accordance with the lunar calendar and thus do not re-occur on the same dates each year.

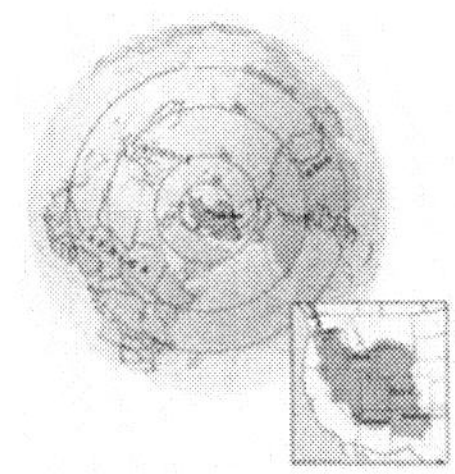

* This is an edited, reformatted and augmented version of a Library of Congress publication dated May 2008.

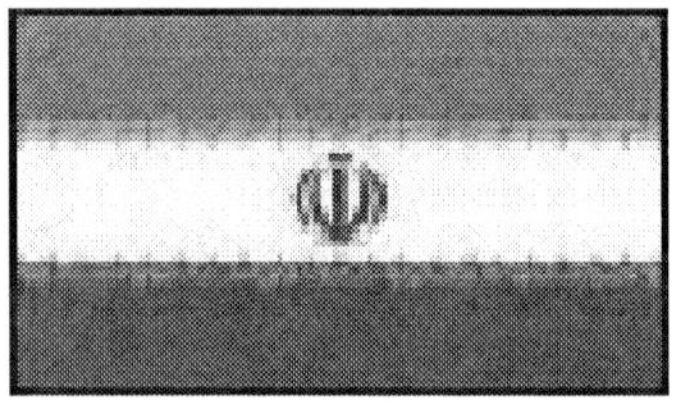

Flag: The flag, adopted after the 1978–79 Revolution, features threehorizontal bars, green (on top), white, and red, representing, respectively, Islam, peace, and courage. The bars are divided by stylized script. Thewhite bar features a centered, red, abstract representation of the nameAllah in the shape of a tulip.

HISTORICAL BACKGROUND

Early History

The first Iranian state was the Achaemenian Empire, established by Cyrus the Great in about 550 B.C. Alexander the Great conquered the empire in 330 B.C. The Greeks were followed by the Parthians, who ruled from 247 B.C. until A.D. 224, and the Sassanians, who ruled from A.D. 224 until the Arabs conquered Iran in A.D. 642. The Arabs brought with them Islam, which eventually became the predominant religion. In the centuries that followed, Iran was ruled by a succession of Arab, Iranian, and Turkic dynasties. In the thirteenth century, the Mongol leader Genghis Khan invaded the disunified territory of Iran, and Mongol dynasties subsequently ruled Iran for nearly two centuries. In 1501 the Iranian Safavis created a strong centralized empire under Ismael I and also established Shia Islam as the official religion. In the eighteenth century, Iran was weakened by civil wars, new dynasties came to rule, and a new regional rival, Russia, arose.

The Qajars and Pahlavis

In 1795 the Qajar family established a dynasty that would rule Iran until 1925. In the nineteenth century, Iran, under the Qajars, lost much of its territory in the Caucasus and Central Asia to Russia. During this period, influence in Iran was divided between Russia and Britain, Russia's chief Western rival in the region. Both powers interfered in local politics and forced Iran to make trade concessions. A popular desire for accountable government and resentment of intrusion by foreign interests led to the Constitutional Revolution of 1905–7 and the formation of a parliament. In 1909 the Anglo-Persian Oil Company was formed and assumed control of Iran's newly discovered oil deposits.

In 1921 army officer Reza Khan provided military support for a coup against the government; he was named minister of defense, then prime minister. Following parliament's deposition of the Qajar dynasty in 1925, he became shah of Iran, adopting the surname Pahlavi. As Reza Shah Pahlavi, he restored order and sought to modernize the economy and society and to forge cultural links abroad. However, in World War II his failure to cooperate with the Allied powers caused Britain and the Soviet Union to invade Iran and force him to abdicate in favor of his son, Mohammed Reza Shah Pahlavi. In 1951 Mohammad Mossadeq, a deputy in the parliament, rode strong Iranian sentiment for nationalization of the oil industry to a position as prime minister. However, in 1953 Britain and the United States, which opposed the principle of oil nationalization at the time, forced the nationalist Mossadeq from power.

In the 1960s, Iran recovered from the economic disruption of the oil nationalization period, but the authoritarian rule of the shah provoked political discontent. It was during this period that Ayatollah Ruhollah Musavi Khomeini developed a following as an antigovernment leader and was sent into exile (1964), first to Turkey and subsequently to Iraq. Between 1965 and 1977, Iran enjoyed improved prosperity, expanded regional influence, and relative stability. However, there was no liberalization of the political system, and repression, coupled with resentment of an increased Western presence, fueled a series of antigovernment protests in 1977 and early 1978 that developed into a nationwide revolutionary movement. From his exile base in Iraq, Khomeini emerged as the leader of an increasingly strong opposition movement, which organized nationwide strikes and effectively paralyzed the economy by late 1978. The military, experiencing mass desertions and the refusal of junior officers to act against strikers and demonstrations, was unable to protect the regime, and the shah decided to leave the country, ostensibly for medical treatment. After the shah's departure, his government was unable to stem what had become a revolutionary tide. Khomeini returned from exile on February 1, 1979, refused to recognize the authority of the shah's prime minister, and appointed a provisional government. When the military announced its neutrality in the power struggle between the two governments, the monarchy effectively ended.

The Islamic Republic of Iran

Following a national plebiscite, an Islamic republic was proclaimed officially on April 1, 1979. The provisional government was composed of a coalition of nationalist and religious leaders who had moderate views with respect to social and economic changes. They were opposed by young militants who advocated radical changes in both domestic and foreign policies. In particular, the latter group wanted to end all ties with the United States.

In November 1979, students affiliated with the latter group occupied the U.S. Embassy in Tehran and held 53 U.S. diplomats hostage for the next 14 months. This incident led to the collapse of the provisional government and a decisive break in U.S.-Iranian relations. Between 1980 and 1988, serious differences between the moderate and militant factions of the revolutionary government were held in check by the need to maintain internal unity during an indecisive war with Iraq that resulted in more than 200,000 Iranian deaths.

In 1989 the death of Khomeini removed the one figure with authority to arbitrate between the two mutually antagonistic political factions of the postrevolutionary elite. Neither of the two factions constituted a homogeneous political group; rather, each comprised multiple ideological tendencies. In general, those whom the Western press labeled "reformists" advocated a liberal interpretation of the constitution and Islamic law but disagreed among themselves with respect to economic, political, and social policies. By contrast, those who came to be known as "conservatives" advocated the strict and literal interpretation of the constitution and Islamic law.

During the presidency of Ali Akbar Hashemi Rafsanjani (1989–97), reformists controlled a majority of seats in parliament until 1992 and supported Rafsanjani's policies for economic reform and the normalization of relations with neighboring countries. The conservatives won a majority of seats in both the 1992 and 1996 parliamentary elections and subsequently used their position in the legislature to weaken or stop outright many reforms proposed by the Rafsanjani government. The administrations of Rafsanjani's successor, Mohammad Khatami (in office 1997–2005), encountered the same resistance. Reformists won a majority of seats in the 2000 parliamentary elections and then enacted several notable pieces of reform legislation in the ensuing term. Having lost control of the parliament, conservatives tried to use their influence in the judiciary and bureaucracy to impede reforms they perceived as threatening their positions. Conservatives regained control of the parliament in the 2004 elections.

Although Iran's foreign relations had improved under Khatami, in the early 2000s earlier progress was eroded by Iran's ostensible support of terrorist groups in the Middle East and conflict with the European Union and the United States over Iran's nuclear program. Iran's international position was jeopardized by the construction of nuclear processing plants to provide fuel for its nuclear energy generating facility at Bushehr, which was being built with significant technical input from Russia. In mid-2005 the surprise election of conservative Mahmoud Ahmadinejad as president led Iran to adopt more confrontational international positions, especially vis-à-vis limitations on its nuclear fuel processing program and its attitude toward Israel. Within Iran, moderate and conservative factions concurred in defending the nuclear program against international interference.

GEOGRAPHY

Location

Iran is located in the Middle East, between Turkey andIraq on the west and Afghanistan and Pakistan on the east; it bordersthe Persian Gulf and Gulf of Oman in the south and Armenia,Azerbaijan, the Caspian Sea, and Turkmenistan in the north.

Size

Iran's total area is nearly 1.65 million square kilometers, of which 1.64 million square kilometers—an area slightly larger than Alaska—is land mass.

Land Boundaries

Iran is bounded by Afghanistan (936 kilometers), Armenia (35 kilometers), Azerbaijan proper (432 kilometers), Azerbaijan's Nakhichevan enclave (179 kilometers), Iraq (1,458 kilometers), Pakistan (909 kilometers), Turkey (499 kilometers), and Turkmenistan (992 kilometers).

Disputed Territory

Iran and the United Arab Emirates (UAE) dispute sovereignty over three islands in the Persian Gulf that are occupied by Iran.

Length of Coastline

Iran's coastline includes 2,440 kilometers on the Persian Gulf and Gulf of Oman and 740 kilometers on the Caspian Sea.

Maritime Claims

Iran's territorial sea extends 12 nautical miles and its contiguous zone, 24 nautical miles. The exclusive economic zone claimed by Iran is determined by bilateral treaties or median lines in the Persian Gulf. The maritime border with Iraq along the Shatt al Arab waterway was established by treaty in 1975 as the median line of the deep-water channel, but the final status of the treaty awaits the conclusion of a peace treaty to replace the 1988 cease-fire agreement between Iran and Iraq. In opposition to at least two of the other four littoral states, Iran advocates a division of the bed of the Caspian Sea that would give Iran control of an area about 20 percent greater than the area it would control under a division based on the actual length of each littoral state's coastline.

Topography

Iran has rugged mountain chains surrounding several basins collectively known as the Central Plateau, which has an average elevation of about 900 meters. East of the Central Plateau are two large desert regions, a salt desert in the north and a rock and sand desert in the south. There are lowland areas along the Caspian coast, in Khuzestan Province at the head of the Persian Gulf, and at several dispersed locations along the Persian Gulf and Gulf of Oman coasts.

Principal Rivers

Iran has no major rivers.The only navigable river is the Karun, which is 830 kilometers long. Other rivers include the Safid Rud (1,000 kilometers), Kharkeh (700 kilometers), and Zayandeh Rud (400 kilometers).

Climate

Iran's climate is mostly arid and semi-arid, with a humid rain-forest zone along the Caspian coast. Temperatures average 10°–25° C in the winter and 19°–38° C in the summer.

Natural Resources

Iran has enormous reserves of oil and natural gas. Oil reserves are estimated at more than 130 billion barrels (third in the world behind Saudi Arabia and Iraq; about 11 percent of world proven reserves), and natural gas reserves are estimated at more than 32 trillion cubic meters (second in the world behind Russia). Mineral resources currently exploited include bauxite, chromium, coal, copper, gold, iron ore, limestone, red oxide, salt, strontium, sulfur, turquoise, and uranium.

Land Use

About 11 percent of Iran's land surface is classified as arable. The most productive agricultural land, bordering the Caspian Sea, makes up about 5.5 percent of the country's total land area.

Environmental Factors

Especially in urban areas, vehicle emissions, refinery operations, and industrial effluents contribute to poor air quality. Between 1985 and 2005, huge increases in energy consumption tripled carbon emissions. Most cars use leaded gas and lack emissions control equipment. Tehran is rated as one of the world's most polluted cities. The abundance of fossil fuels has discouraged use of alternative fuels. Much of Iran's territory suffers from desertification and/or deforestation. Industrial and urban wastewater runoff has contaminated rivers and coastal waters and threatened drinking water supplies. Wetlands and bodies of freshwater increasingly are being destroyed as industry and agriculture expand, and oil and chemical spills have harmed aquatic life in the Persian Gulf and the Caspian Sea. Iran contends that the international rush to develop oil and gas reserves in the Caspian Sea presents that region with a new set of environmental threats. Although a Department of Environment has existed since 1971, Iran has not developed a policy of sustainable development because short-term economic goals, particularly support of the oil and gas industries, have taken precedence.

Time Zone

Iran lies in one time zone, which is three and one-half hours ahead of Greenwich Mean Time.

SOCIETY

Population

Iran's population is about 70 million according to preliminary data from the decennial census conducted in late 2006; of that number, approximately one-third is rural and two-thirds urban. Urbanization has been steady; in 1976 only 47 percent of the population lived in urban areas. Population density averages 42 people per square kilometer, but with significant regional variations. In 2008 the estimated annual population growth rate was less than 1 percent (0.79 percent). Net migration in 2008 was an estimated –3.28 persons per 1,000 population. In 2006 Iran hosted more than 660,000 Afghan and 54,000 Iraqi refugees.

Demography

According to a 2008 estimate, 22.3 percent of Iran's population is 14 years of age or younger, and only 5.4 percent is 65 and older. The median age is 26.4 years. There are 1.03 males for every female. Estimated life expectancy is 70.86 years overall (69.39 years for men, 72.4 years for women). The birthrate is 16.89 per 1,000; the death rate, 5.69 per 1,000; and the infant mortality rate, 36.73 per 1,000 live births. The fertility rate remains at about 1.7 children born per woman, a significant reduction from the estimated rate of 7.0 in 1979.

Ethnic Groups and Languages

The main ethnic groups in Iran are Persians (65 percent), Azerbaijani Turks (16 percent), Kurds (7 percent), Lurs (6 percent), Arabs (2 percent), Baluchis (2 percent), Turkmens (1 percent), Turkish tribal groups such as the Qashqai (1 percent), and non-Persian, non-Turkic groups such as Armenians, Assyrians, and Georgians (less than 1 percent). Persian, the official language, is spoken as a mother tongue by at least 65 percent of the population and as a second language by a large proportion of the remaining 35 percent. Other languages in use are Azeri Turkish and Turkic dialects, Kurdish, Luri, Arabic, and Baluchi. Azeri and Kurdish autonomy movements arose in the 1940s, and a Kurdish autonomy movement was active in the period 1979–83.

Religion

The constitution declares Shia Islam to be the official religion of Iran. At least 90 percent of Iranians are Shia Muslims, and about 8 percent are Sunni Muslims. Other religions present in Iran are Christianity (mainly Armenians and Assyrians, more than 300,000 followers), the Baha'i faith (at least 250,000), Zoroastrianism (about 32,000), and Judaism (about 30,000). The constitution recognizes Christianity, Judaism, and Zoroastrianism as legitimate minority religions. The Baha'i faith is not recognized as a legitimate minority religion, and since 1979 Baha'is have experienced periodic bouts of persecution. Individuals of all religions are required to observe Islamic codes on dress and gender segregation in public. Individuals of minority religions are prohibited from serving in senior administrative positions in many government ministries. In the early 2000s, Christians have been emigrating from Iran at the rate of 15,000 to 20,000 per year.

Education and Literacy

In 2003 the literacy rate of the population was 79.4 percent (85.6 percent for males and 73 percent for females). Under the constitution, primary education (between ages six and 10) is compulsory, and primary enrollment was nearly 98 percent in 2004. Secondary school attendance is not compulsory. Hence, enrollment rates are lower—about 90 percent for middle school and 70 percent for high school in 2004. Primary, secondary, and higher education is free, although private schools and universities charge tuition. Minority religions except the Baha'i faith maintain private schools, but supervisors must be Muslim, and one hour per week of Islam is a required subject, as in public schools. The majority of Iran's 113,000 precollegiate public schools are single-sex beyond kindergarten. Universities are coeducational. By 2004, Iran had more than 200 public and more than 30 private institutions of higher education, enrolling a total of nearly 1.6 million students. The largest and most prestigious public university is the University of Tehran. The largest private university is the Islamic Free University.

Health

The overall quality of public health care improved dramatically after the 1978–79 Revolution because public health has been a top priority of the government. The constitution entitles Iranians to basic health care, and most receive subsidized prescription drugs and vaccinations. An extensive network of public clinics offers basic care at low cost, and general and specialty hospitals operated by the Ministry of Health provide higher levels of care. In most large cities, well-to-do persons use private clinics and hospitals that charge high fees. Specialized medical facilities are concentrated in urban areas, but rural communities have relatively good access to primary care physicians at clinics in villages, where the government-sponsored primary health care system has raised the level of health education and prenatal care since the late 1990s. Immunization of children is accessible to most of the urban and rural population. In the early 2000s, estimates of the number of physicians varied from 8.5 to 11 per 10,000 population. About 46 percent of physicians were women. There were about seven nurses and 11 hospital beds per 10,000 population. Some 650 hospitals were in operation. In the early 2000s, about 65 percent of the population was covered by the voluntary national health insurance system. More expensive private health insurance plans also were available.

As Iran's health system has improved, the role of communicable diseases as causes of death has diminished relative to that of noncommunicable diseases. Therefore, in the early 2000s the main natural causes of death have been cardiovascular disease and cancer. Opium and other drug addictions constitute a major and growing health problem; in 2005 estimates of the number of drug addicts ranged from 2 to 4 million. Increased drug use has driven up the incidence of human immunodeficiency virus (HIV). The HIV/acquired immune deficiency syndrome (AIDS) prevalence rate in Iran was estimated at 0.2 percent in 2005. An estimated 66,000 people were living with HIV/AIDS, and 1,600 deaths were attributed to HIV/AIDS. Iran has established a national HIV treatment system, including 150 testing sites and a free needle exchange program. Earthquakes regularly take several thousand lives in Iran. The Bam earthquake of December 2003 killed nearly 28,000 people.

Welfare

Iran's Ministry of Social Affairs supervises public programs for pensions, disability benefits, and income for minor children of deceased workers. Welfare programs for the needy are managed by more than 30 individual public agencies and semi-state organizations, as well as by several private nongovernmental organizations. In 2003 the government began to consolidate its welfare organizations in an effort to eliminate redundancy and inefficiency. The largest welfare organization is the Bonyad-e Mostazafin (Foundation of the Disinherited), a semi-public foundation originally founded in 1979 with the assets of the last shah's family; it operates a wide variety of charitable activities. In late 2005, President Ahmadinejad formed the Reza Love Fund to provide financial assistance to young couples seeking financial stability.

In the late 1990s, the extension of pensions to farming household heads over 60 effectively doubled the number of Iranians eligible for government pensions to more than 60 percent of the workforce. Self-employed persons in urban areas are the major group not covered. Civil servants, the regular military, law enforcement agencies, and the Islamic Revolutionary Guard Corps have their own pension systems. In 2003 the minimum standard pension was 50 percent of the worker's earnings but not less than the amount of the minimum wage. Iran spent 22.5 percent of its 2003 national budget on social welfare programs. More than 50 percent of that amount covered pensions. Considering all social welfare programs available, urban residents benefit more than the rural population. Government workers are eligible for sickness, maternity, and work injury benefits, but few private employers provide these benefits. The Imam Khomeini Social Assistance Committee and other semi-public foundations supply such benefits to some workers.

Economy

Overview

Iran's economy is dominated by the oil industry, which is part of the state sector. In the early 2000s, more than 80 percent of export earnings came from oil and gas. The state also owns and administers several large industries. The private sector includes automobile, textile, metal manufacturing, and food-processing factories as well as thousands of small-scale enterprises such as workshops and farms. Smuggling and other illegal economic activities occupy an increasingly large part of the overall economy. The economic reform programs of presidents Ali Akbar Hashemi Rafsanjani (in office 1989–97) and Mohammad Khatami (in office 1997–2005) aimed at making Iranian industry more competitive internationally. Measures included selling off government enterprises, reducing subsidies, creating an equitable income tax system, and cutting high tariffs that protect local manufacturing from foreign imports. Private business interests strongly opposed many reforms and were able to block their enactment. A dominant share of Iran's nonpetroleum industrial output is controlled by semi-private charitable organizations called *bonyads*, which exert considerable influence on economic policy through their close links to powerful politicians. Traditional import-export merchants, collectively known as the bazaar, also occupy an influential place in economic policy making. Because export earnings from oil and gas accrue to the government as revenue, world prices for those commodities have a major impact on Iran's budget. Government economic planning is based on five-year development plans, the fourth of which began in March 2005. Although economic diversification has been a goal in the early 2000s, little progress has been made in that direction.

Gross Domestic Product (GDP)

In 2007 Iran's GDP (at the official exchange rate) was estimated at US$278.1 billion; GDP in terms of purchasing power parity (PPP) was US$852.6 billion, or US$12,300 per capita. The estimated GDP growth rate was 4.3 percent. Industry contributed an estimated 45.3 percent, services 43.7 percent, and agriculture 11 percent of GDP.

Government Budget

Budget deficits, largely caused by extensive state subsidies of commodities such as food and fuel, have been a major problem. Iran's 2005 budget included a deficit of US$11.6 billion based on revenues of US$48.8 billion and expenditures of US$60.4 billion, including US$7.6 billion of capital expenditures. Backed by increased oil prices, announced outlays in the 2006 budget were about 50 percent more than those in 2005.

Inflation

The government's anti-inflationary policies have reduced inflation from the average rate of 23 percent in 1977–98. The official rate was 15.8 percent in 2006 and an estimated 17 percent in 2007.

Agriculture

Iran's diversity of terrain and climate enable cultivation of a variety of crops, but in 1998–2000 severe droughts cut agricultural production. Output has recovered slowly since, although many villages in eastern Iran have been abandoned, and consequently the area under cultivation has decreased since 2000. In the early 2000s, about 20 percent of Iran's arable land remained uncultivated. Iran is a net importer of grains, especially rice and wheat, and a net exporter of fruits, nuts, and various specialty crops. Iranians grow a variety of crops, most notably wheat, barley, rice, pistachio nuts, cotton, sugar beets, and sugarcane. About one-third of agricultural income comes from livestock, chiefly chickens, sheep, beef cattle, and dairy cows. With the exception of sheep and goats, which graze on open rangeland, most livestock is raised in fenced pastures.

Forestry

Iran has only about 7 percent forest cover. The major commercially useful forests are located in the Alborz Mountains in the north, especially on the southern slopes above the Caspian Sea coast. Smaller forests, principally of oak and other deciduous trees, are scattered throughout the Zagros Mountains in western and central Iran. Iran is a net importer of timber products. Illegal clear-cutting and clearing for agriculture have depleted forests in the Alborz, and illegal harvesting of trees has hindered government replanting programs. In 2003 the timber industry produced about 13 million cubic meters of wood products, of which about 37 percent was pulpwood and 24 percent logs.

Fishing

Iran has a long tradition of fishing in the Caspian Sea, in the Persian Gulf, and on inland rivers. The government company Sheelat establishes fishing quotas and buys fish for processing. Most of the actual fishing is undertaken by small-scale, private fishermen. The most economically important product of the fishing industry is caviar from Caspian Sea sturgeon. In the 1990s, the sturgeon catch declined as a result of over-fishing and poaching. Iran has an aggressive fish nurseries program aimed at reversing the decline in Caspian fish stocks. Other products of the fishing industry are tuna, the sardine-like kilka, trout, and shrimp. In 2004 catches totaled more than 330,000 tons of fish.

Mining and Minerals

In 2006 Iran produced more than 15 different nonradioactive metals and 27 nonmetal minerals. The mined products yielding the greatest value were iron ore, decorative stones, gravel and sand, coal, copper ore, and limestone. The fastest growing nonpetroleum extraction industry is copper. Iran has an estimated 4 percent of the world total of copper.

Industry and Manufacturing

Iran's most important industries are those associated with the extraction and processing of oil and gas. Iran's manufacturing output was reduced during the 1978–79 Revolution, but the 1980–88 war with Iraq had the indirect consequence of increasing industrial production. In the 1990s, low private investment levels hindered growth, although government expenditures based on revenues from high world oil prices stimulated public investment and also directly stimulated consumer demand and the petrochemicals industry. The petrochemicals industry has grown rapidly in the early 2000s; the Fourth Economic Development Plan (2005–10) calls for a major expansion of annual petrochemical output, from 9 million tons in 2001 to 27 million tons by 2015.The steel industry, centere d in Ahvaz, Esfahan, and Mobarakeh, also has grown rapidly since 1990. The output goal for 2004 was 8.5 million tons. Automobile manufacture has benefited from licensing agreements with European and Asian manufacturers. In 2002 the largest plant, Iran Khodro, built about 260,000 units, and several smaller facilities produced a total of about 240,000 vehicles. Processing of agricultural products also is an important industry and is dominated by domestic private firms. Among the major subsectors are grain processing and fruit and vegetable canning. The textile industry, based on domestic cotton and wool, employed about 400,000 people in 2000. The construction industry has grown rapidly since 2000 because of government investment in infrastructure projects and increased demand for private housing. Throughout history, Iranians have been involved in handicraft activities, producing a variety of products, most notably carpets.

Energy

Iran possesses abundant fuels from which to generate energy, ranking second in the world in natural gas reserves and third in oil reserves. Nevertheless, in 2005 Iran spent US$4 billion dollars on fuel imports, mainly because of inefficient domestic use. Oil industry output averaged about 4 million barrels per day in 2005 and 2006, compared with the peak output of 6.6 million barrels per day reached in 1976. In the early 2000s, industry infrastructure was increasingly inefficient because of technological lags.

Natural gas output in 2006 was 105 billion cubic meters. A large share of Iran's natural gas reserves are believed to remain untapped, although gas already accounted for nearly one-half of energy consumption. With massive government investments planned, the share of gas in energy production was expected to rise quickly in ensuing years.

By 2004 the addition of new hydroelectric stations and the streamlining of conventional coal- and oil-fired stations increased installed capacity to more than 39,000 megawatts. Of that amount, in 2006 about 50 percent was based on natural gas, 18 percent on oil, and 6 percent on hydroelectric power. In 2004 Iran opened its first wind-powered and geothermal plants, and the first solar thermal plant was to come online in 2009. Demographic trends and intensified industrialization have caused electric power demand to grow, averaging 5 percent per year from 1967 to 2004 and projected to average 7 percent per year from 2005 to 2010. The government's goal of 53,000 megawatts of installed capacity by 2010 is to be reached by building new gas-fired plants financed by independent power producers, including those with foreign investment backing, and by adding hydroelectric and nuclear power generating capacity. Iran's first nuclear power plant at Bushehr, which may be completed in 2008 after years of delays, has received international criticism because of concerns that its enriched uranium and spent fuel can be diverted for the production of nuclear weapons.

Services

In the financial sector, Iran's banking system was nationalized in 1979. Private banks were not authorized to reopen until the early 2000s, when some small private credit institutions appeared. All state and private banks are strictly overseen by the Central Bank of Iran, also known as Bank Markazi. Accounts of the state-owned commercial banks are dominated by loans to state and *bonyad* enterprises and to large-scale private firms. Wealthy Iranians use foreign banks, especially for savings accounts. The Fourth Economic Development Plan (2005–10) calls for the introduction of foreign banks, but such a move has met with substantial resistance.

The trading of shares on the Tehran Stock Exchange was limited between 1979 and 1986, but activity has increased sharply since 2002. In 2007 some 364 companies were listed on the exchange, up from 164 in 1996.

The tourism industry was disrupted by the 1978–79 Revolution and the Iran–Iraq War (1980–88) but began to revive in the 1990s. In FY 2004 some 700,000 foreign tourists visited Iran. The Fourth Economic Development Plan calls for the number of foreign tourists to increase to 2.5 million annually. However, the industry faces serious limitations in infrastructure, communications, regulatory norms, and personnel training.

Labor

In 2007 Iran's labor force totaled 28.7 million. An estimated 14 percent of the labor force was unemployed; the unemployment rate was much higher among younger workers. Underemployment also was common. The Fourth Economic Development Plan, which began in 2005, aimed to create 700,000 new jobs per year, but unemployment remained unchanged during the first year of that plan. Skilled labor has been in short supply. In 2007 about 45 percent of the labor force was employed in services, 31 percent in industry, and 25 percent in agriculture. In 2005 the minimum wage was about US$120 per month. That level provoked substantial labor unrest in 2005.

Foreign Economic Relations

The Iran Sanctions Act, in existence since 1996 and until 2006 known as the Iran–Libya Sanctions Act, is a full U.S. trade embargo against Iran, although the embargo was relaxed in 2000 to permit U.S. companies to import Iranian carpets, caviar, and pistachio nuts. Other countries, including members of the European Union, have continued to trade with Iran, but Western countries have blocked the export to Iran of dual-use items, such as computer equipment, with potential military applications. In the early 2000s, China emerged as an important trade partner in both imports and exports. Japan retained the position that it assumed in the mid-1990s as Iran's best export customer. In order of volume, the main purchasers of Iran's exports in 2006 were Japan, China, Turkey, Italy, South Korea, and the Netherlands. In order of volume, the main source countries for Iran's imports in 2006 were Germany, China, the United Arab Emirates, France, Italy, South Korea, and Russia. The main commodities imported are basic manufactures, chemicals, food (chiefly rice and wheat), and machinery and transport equipment. The main commodities exported are petroleum, carpets, chemical and petrochemical products, fruit and nuts, iron and steel, natural gas, and copper.

Trade Balance

In 2007 Iran's estimated income from exports was US$76.5 billion (free on board—f.o.b.), 85 percent of which came from petroleum and natural gas. The estimated payment for imports in 2007 was US$61.3 billion (f.o.b.), yielding a trade surplus of US$15.2 billion.

Balance of Payments

In 2007 Iran's current account balance, determined mainly by its merchandise trade surplus and its smaller services trade deficit, was US$19 billion. Its foreign exchange reserves, determined primarily by oil prices, were estimated at US$69.2 billion in 2007. Records on portfolio investment are not available. Foreign direct investment has remained relatively small.

External Debt

In 1991 Iran's external debt was estimated at US$23 billion. During the next decade, Iran paid down the debt, which dropped to US$7.8 billion in 2001. Subsequently, the debt has risen as international borrowing has increased. The 2006 estimate was US$18.6 billion and the 2007 estimate, $13.8 billion.

Foreign Investment

Foreign investment has been hindered by unfavorable or complex operating requirements in Iran and by international sanctions, although in the early 2000s the Iranian government liberalized investment regulations. In the early 2000s, foreign investors have concentrated their activity in a few sectors of the economy: the oil and gas industry, vehicle manufacture, copper mining, petrochemicals, foods, and pharmaceuticals. Iran's constitution prohibits direct concession of petroleum rights to foreign investors. In the 1990s and early 2000s, some indirect oilfield development agreements were made with foreign firms. The most active investors have been British, French, Japanese, South Korean, Swedish, and Swiss companies.

Currency and Exchange Rate

The value of the rial, Iran's unit of currency, declined substantially between 2002 and 2005. In 2002 a multiple exchange rate was replaced by a single floating rate. The average exchange rate per US$1 was 9,227.1 rials in 2006 and 9,407.5 rials in 2007. In mid-May 2008, the exchange rate was approximately 9,255 rials to the U.S. dollar. The tuman, which is worth 10 rials, is the preferred unit of currency in commerce.

Fiscal Year

In accordance with the Iranian calendar year, the fiscal year begins March 21 and runs through March 20.

Transportation and Telecommunications

Overview

Constant construction and expansion of the road and rail networks, even during the Iran–Iraq War (1980–88), have resulted in an overland transportation system that is adequate for freight and passenger demands. Ports destroyed during the war have been rebuilt, and new ones on the Caspian Sea and Persian Gulf have been developed. Air transportation is relatively inexpensive, and all large cities and many smaller ones have airports with regularly scheduled daily flights.

Roads

Under Reza Shah Pahlavi (ruled 1925–41), a construction program provided Iran with a national road system, which expanded in the 1960s and 1970s. After the 1978–79 Revolution, road construction programs focused on connecting rural areas to provincial cities. Since 1989, road construction has stressed ring roads around large cities and multilane highways between major metropolitan areas. In 2003 Iran had a total of 100,000 kilometers of paved roads and nearly 80,000 kilometers of graded, unpaved roads. The three national auto routes are the A–1 across northern Iran from the Turkish border in the west to the Afghan border in the east, and connecting Tabriz, Tehran, and Mashhad; the A–2 across southern Iran from the Iraqi border in the west to the Pakistani border in the east; and the Tehran–Qom–Esfahan–Shiraz highway, which traverses central Iran from north to south. Beginning in the mid-1990s, subsidized gasoline supplies and increased domestic automobile manufactures have spurred overcrowding of the road system.

Railroads

The rail system, which originally was constructed in the 1920s and 1930s, has been undergoing constant expansion since 1989. In 2006 Iran had a total of 8,367 kilometers of rail lines. Only 13 of Iran's 30 provinces had railroad service in the early 2000s. The five main lines of the national system radiate from Tehran: one runs south to Khorramshahr and Abadan at the head of the Persian Gulf; one runs south to the Strait of Hormuz at Bandar-e Abbas; one runs southeast to Kerman, with a route under construction farther east to Zahedan, which already is connected to the Pakistan State Railways; one runs east to Mashhad, connecting with the Central Asian system on the Turkmenistan border and including a spur to the eastern side of the Caspian Sea; and the fifth line runs northwest to Tabriz and the border with Turkey, where it connects to the Turkish State Railroad and includes a spur north of Tabriz to Azerbaijan's Nakhichevan enclave. Tehran also has a combined underground and surface rail commuter system, and metro rail systems are under construction in Esfahan and Shiraz.

Ports

In 2004 about 53 million tons of cargo were unloaded and 30 million tons loaded at Iran's 14 ports. More than one-third of total traffic came through Bandar-e Abbas on the Strait of Hormuz. The main oil terminal is at Khark Island, located 25 kilometers offshore in the northeastern Persian Gulf. Since 1992, Caspian ports have handled more trade as commerce with the Central Asian countries has increased. Modernization projects are underway at Bandar-e Anzali on the Caspian Sea and at Chabahar on the Gulf of Oman. The national Islamic Republic of Iran Shipping Lines (IRISL) has routes in the Persian Gulf and Caspian Sea and carries cargo to Europe and the Far East.

Inland Waterways

In 2006 Iran had 850 kilometers of inland waterways. The most important is the 193-kilometer-long Shatt al Arab (Arvanrud in Persian), which is formed in Iraq by the confluence of the Tigris and Euphrates rivers. The Shatt al Arab then forms the Iran–Iraq border until it flows into the Persian Gulf. The ports of Abadan and Khorramshahr are located along the Shatt al Arab.

Civil Aviation and Airports

In 2007 Iran had 331 airports, 129 of which had paved runways. Of those, 41 (40 paved and one unpaved) had runways 3,000 meters or longer. International airports are located at Tehran, Tabriz, Mashhad, Bandar-e Abbas, Bushehr, Esfahan, and Shiraz, and on the islands of Kish in the Persian Gulf and Qeshm in the Strait of Hormuz. In 2004 Iran's airports served about 4.3 million international passengers, about two-thirds of whom flew on domestic airlines. Imam Khomeini International Airport in Tehran has an annual capacity of 6.5 million passengers, but its final design capacity is 40 million passengers and 700,000 tons of cargo. Plans call for a gradual expansion of services. In 2006 the national airline, Iran Air, served 25 cities in Iran with connections to the Persian Gulf and European and Asian cities; it operated a fleet of 36 aircraft and employed about 12,000 workers. The second-largest carrier, the private Asseman Airlines, connects the largest domestic cities with destinations on the Persian Gulf and elsewhere in Asia. In 2007 some 14 heliports also were in operation.

Pipelines

In 2007 Iran had 19,161 kilometers of natural gas pipelines, 8,438 kilometers of oil pipelines, 7,936 kilometers of pipelines for refined products, 570 kilometers of pipelines for liquid petroleum gas, and 397 kilometers of pipelines for gas condensate. Iran's central pipeline infrastructure is designed for the distribution of natural gas for domestic use and for the domestic transit of oil, including from offshore oil fields to processing centers. That structure has been supplemented as the natural gas industry and the fuel export industry expanded. Since 2000 several new natural gas pipelines have been planned; some have failed because of geopolitical considerations (for example, U.S. opposition to a key Iranian role in delivering Central Asian oil and gas to the West), and some, such as the gas export line from Iran to Turkey, function at reduced capacity. In 2006 plans called for new pipelines to exploit markets in Armenia and Pakistan. In 2007 a new 160-kilometer line to Armenia began operations. However, a 2,600-kilometer line to Pakistan, which potentially also could supply India, remains in the negotiation stage.

Telecommunications

Most phases of telecommunications services are controlled by the state. Between 1995 and 2006, the number of telephone landlines increased from 86 to 330 per 1,000 population. A large-scale modernization program, backed by heavy state investment, has aimed at improving and expanding urban service and reaching rural areas that lack telephone service. In 2006 Iran had nearly 22 million main telephone lines in use, and an estimated 13.7 million subscribers had mobile telephone service. Internet use expanded rapidly in the early 2000s from an estimated 250,000 users in 1999 to an estimated 7.5 million users in 2005. The Telecommunications Company of Iran estimated that Iran would have 25 million Internet users by 2009. However, the state filters Internet content intensively.

GOVERNMENT AND POLITICS

Political System

Following the Islamic Revolution of 1978–79, a national referendum approved a new constitution; several amendments were approved in 1989. According to that constitution, the Islamic Republic of Iran is a republic with nominal separation of powers among the executive, judicial, and legislative branches. The senior figure in the system is the *faqih* (leader*),* an expert in religious law, who is referred to in the constitution as the Leader of the Revolution. The constitution named Ayatollah Khomeini as the first *faqih* by virtue of his leading role in the 1978–79 Revolution. Leaders are elected by a majority vote of the Assembly of Experts, a body of senior clergymen who are elected in national elections. The Assembly of Experts elected then president Sayyid Ali Khamenei to succeed Khomeini in June 1989. The legal system is based on sharia (Islamic law).

Executive Branch

The leader, who exercises many de facto executive functions,is elected by a majority vote of the Assembly of Experts, an 86-member body of senior clergymen who are elected by popular vote to eight-year terms. The Assembly evaluates the work of the leaderin annual meetings; it can dismiss the leader if he is deemed no longer qualified. The leaderis responsible for choosing the commanders of the military services and the head of the judiciary, setting general state policy, declaring war and peace, commanding the armed forces (including control of intelligence and security agencies), initiating and supervising amendments to the constitution, and supervising a variety of influential parastatal foundations and organizations. The executive branch is headed by the president, who in practice is the second-highest government official. He is elected in national elections every four years and is limited to two consecutive terms. The constitution specifies that the president must be a Shia Muslim. The current president, Mahmoud Ahmadinejad, was elected in 2005. The president selects several vice presidents (10 listed in 2008) and the 21 ministers who constitute his cabinet. Ministers but not vice presidents are subject to approval by parliament. The leader can dismiss a president if two-thirds of parliament votes to impeach him.

The relationship between the president and the leader, not well defined by the constitution, has varied with the individuals in power. The strong personality of Ali Akbar Hashemi Rafsanjani (president, 1989–97) made him the most authoritative person in the political system, and as leaderKhamenei acquiesced to his policies. By c ontrast, the reluctance of Rafsanjani's successor, Mohammad Khatami, to engage in confrontational politics enabled Khatami's conservative opponents to advance the authority of the leader as superior to that of the president. In his early presidency, Ahmadinejad, whom Khamenei did not back in the first round of the 2005 presidential election, presented himself as a restorer of the revolutionary ideals of Ayatollah Khomeini, implicitly criticizing Khamenei and the other religious leaders of recent years.

Legislative Branch

The legislative branch consists of a parliament, or Majlis, and theGuardians Council. The Majlis comprises 290 deputies who are elected to four-year terms on the basis of universal suffrage. Five of these seats are reserved for special representatives of officially recognized religious minorities: two for Armenian Christians and one each for Assyrian Christians, Jews, and Zoroastrians. In 2005 some 12 women held seats. The speaker presides over parliament, assisted by two deputies and a system of 22 permanent committees. Select committees also can be established when necessary.

The Majlis may both propose and pass legislation, and the executive branch cannot dissolve it. Ministers of the cabinet can also present bills. All bills passed by the Majlis must be reviewed by the 12-member Guardians Council for consistency with the constitution and with Islamic principles. Members of the Guardians Council must be lawyers; the leader and the Majlis each appoint six members. If the Guardians Council finds a bill compatible with the constitution and Islam, the bill becomes law; if it finds a bill partially or wholly unconstitutional or un-Islamic, the bill is sent back to the Majlis for revision. In 1987 Khomeini resolved tension that had developed between the Majlis and the Guardians Council by establishing the Expediency Council to resolve disputes between the two. In practice, the Expediency Council has upheld some Guardians Council vetoes, overridden others, and sent back some vetoed legislation with instructions that the Majlis and Guardians Council work out acceptable compromises. This pattern continued in both the Rafsanjani and Khatami administrations.

Judicial Branch

The highest judicial authority is the Supreme Court, members of which are appointed by the head of the judiciary. That individual, appointed to a five-year term by the leader, also approves the candidate list from which the president chooses a minister of justice. The Supreme Court nominally has 33 regional branches, to which the chief of the Supreme Court assigns cases, but all but two are located in Tehran. The Supreme Court oversees enforcement of the laws by lower courts, sets judicial precedent, and acts as a court of appeal. Public courts try conventional civil and criminal cases at the provincial and local levels. Revolutionary courts try cases involving political offenses and national security. The Clerical Court, which is outside the court system and overseen directly by the leader, deals with crimes committed by members of the clergy, including "ideological offenses." Such offenses include interpretations of religious precepts that are not acceptable to the establishment clergy and activities, such as journalism, outside the realm of religion. Iran also has special courts for members of the security forces and government officials. The judges of all courts must be experts in Islamic law.

Administrative Divisions

Iran is divided into 30 provinces. The provinces are subdivided into counties (321 in 2007), districts, and villages.

Provincial and Local Government

Each province is administered by a governor general appointed by the central government. The governor general, in consultation with the Ministry of Interior, then appoints the governor of each county in the province and, in consultation with the latter, the chief of each district. At the local level, directly elected city and village councils have exerted substantial authority since the first local elections in 1999. Conservative candidates swept most of the local council elections held in 2003, but moderates won the majority of local council seats in the 2007 elections.

Judicial and Legal System

Although the constitution provides for an independent judiciary, in practice the judicial branch is influenced strongly by political and religious institutions. Defendants have the right to public trial, choice of a lawyer, and appeal. Judicial authority is concentrated in the judge, who also acts as prosecutor and investigator with no legal counsel. In the early 2000s, reformers tried unsuccessfully to gain Majlis approval for the introduction of jury trials. The Islamic revolutionary courts have authority to hold suspects for long pre-trial periods without the benefit of counsel. Charges often are vague, such as "anti-state activity" or "warring against God," and lawyers have complained of being harassed and even imprisoned.

Electoral System

Suffrage is universal at age 16. Direct elections every four years choose the Majlis, president, and local councils. Because these elections are not held simultaneously, Iranians generally vote in a national election every year. Each of the 290 seats of the Majlis nominally represents constituencies of about 200,000, but distribution favors urban areas. The city of Tehran, for example, has 30 at-large constituencies. Candidates for office at any level may simply declare themselves by filing a registration form and paying a nominal fee. The Ministry of Interior and the Central Oversight Committee of the Guardians Council vet candidates for the presidency, parliament, and Assembly of Experts. Local boards supervise elections at the lowest governmental levels. Important qualifications for candidacy are a history of participating in the 1978–79 Revolution and a reputation for being a devout Muslim and observer of Islamic law. Postsecondary education also is relevant for national office. Candidates for the Assembly of Experts must be senior Islamic clergymen. The Guardians Council has used its vetting capacity to disqualify a high percentage of reformist candidates.

Politics and Political Parties

Political parties were legalized in 1998. However, official political activity is permitted only to groups that accept the principle of political rule known as *velayat-efaqih*, literally, the guardianship of the *faqih* (religious jurist). Allegiances, still based on special interests and patronage, remain fluid. In 1998, 18 parties joined in a broad coalition called the Second of Khordad coalition. All were reformist parties that supported the political and economic proposals of President Mohammad Khatami; in the early 2000s, internal differences over specific economic policies have hampered the coalition's effectiveness, however. During that period, the conservatives were more united, despite the existence of several major conservative parties. The Islamic Iran Builders Council (known as Abadgaran) emerged as a powerful conservative coalition beginning in 2003, leading the conservatives to victory in the 2004 parliamentary elections and the 2005 presidential election. Conservatives also prevailed in the 2008 parliamentary elections.

Mass Media

The constitution guarantees freedom of the press, provided that published material complies with Islamic principles. Freedom of speech is not guaranteed. In 1997 and 1998, the government relaxed regulations for publishing licenses, which led to the emergence of scores of new, mostly reformist newspapers and journals. Conservatives reacted by suing individual papers and publishers for libel and in the year 2000 succeeded in forcing the suspension, temporary or permanent, of several dozen newspapers. Nevertheless, the judiciary generally allows some reformist publications to remain open at any given time, and reformist publications continue to express views on many contentious issues. The newspapers with the largest circulation are published in Tehran and include the conservative *Jomhuri-e Islami* (Islamic Republic), *Kayhan* (World), and *Resalat* (Prophetic Mission). The state news service is the Islamic Republic News Agency, which publishes the English-language*Iran Daily*. Several foreign news agencies maintain offices in Tehran, including Agence France Press, Anadolu Ajansi, ITAR–TASS, Reuters, and Tsinhua.

Radio and television broadcasting is controlled by the state. Iran was reported to have 28 television stations (450 repeaters), as well as 72 AM, five FM, and two short-wave radio stations as of the late 1990s. In 2002 an estimated 13 million radios and 7 million television sets were in use. Although they are officially forbidden, and jamming has occurred, an estimated 1 million satellite dishes were used to receive international broadcasts. In 2003 the government assigned a commission to monitor Internet news sites.

Foreign Relations

The election of Mohammad Khatami in 1997 led to improved relations with Iran's neighbors and with most of the West, excluding Israel and the United States. The Khatami government (in power 1997–2005) stressed commercial and geopolitical relations with Western Europe and Japan. Foreign relations have been an area of consensus among conservatives and reformers since the late 1980s. In the early 2000s, attempts by the Khatami regime to find common ground with the United States did not achieve the desired normalization of bilateral relations. In fact, the Bush administration's inclusion of Iran as part of an "axis of evil" in 2002 brought relations to a new low in the post-1989 period. Beginning in 2004, relations deteriorated further because U.S. officials believe that Iran intends to develop nuclear weapons. The Bush administration also expressed concern that Iran might be supporting insurgents against U.S. forces in Iraq and Palestinian groups opposed to the peace process in Israel. As relations with the United States remained strained and mutually distrustful, relations with Europe also declined because of Iran's insistence on processing nuclear fuel for its nuclear energy plant. Russia, which since 1995 has been building Iran's twin nuclear energy reactors for electricity generation, tried to reach a compromise agreement whereby Iran would agree to transfer all of its nuclear fuel processing program to Russia.

Since the overthrow of Iraqi President Saddam Hussein in 2003, Iran has established cooperative relations with the interim government, which is dominated by Iraqi Shia political parties that have had a close relationship with Iran since 1979. In the early 2000s, relations with other regional Arab countries have varied from "correct" or relatively good (e.g., with Jordan, Kuwait, Oman, Qatar, Saudi Arabia, Syria, and the United Arab Emirates) to relatively strained (e.g., with Algeria and Egypt). Iran also has had relatively good relations with China, India, and Russia, particularly in the area of military cooperation. Relations with neighbors Pakistan and Turkey have been correct but not close; in 2006 a possible gas pipeline deal had the potential to improve relations with Pakistan.

Membership in International Organizations

Iran is a member of the following international organizations: Colombo Plan, Economic Cooperation Organization, Food and Agriculture Organization, Group of 15, Group of 24, Group of 77, International Atomic Energy Agency, International Bank for Reconstruction and Development, International Civil Aviation Organization, International Control Commission, International Criminal Police Organization (Interpol), International Development Association, International Development Bank, International Federation of Red Cross and Red Crescent Societies, International Finance Corporation, International Fund for Agricultural Development, International Labour Organization, International Monetary Fund, International Organization for Migration, International Telecommunication Union, Islamic Development Bank, Multilateral Investment Guarantee Agency, Organisation for the Prohibition of Chemical Weapons, Organization of the Petroleum Exporting Countries, Pollution Control Agency, United Nations, United Nations Committee on Trade and Development, United Nations Educational, Scientific and Cultural Organization, United Nations High Commissioner for Refugees, United Nations Industrial Development Organization, Universal Postal Union, World Confederation of Labor, World Federation of Trade Unions, World Health Organization, World Tourism Organization, and World Trade Organization (observer status).

Major International Treaties

Iran is a signatory to international environment agreements on Biodiversity, Climate Change, Climate Change–Kyoto Protocol, Desertification, Endangered Species, Hazardous Wastes, Marine Dumping, Ozone Layer Protection, Ship Pollution, Wetlands (signed but not ratified), Environmental Modification, Law of the Sea, and Marine Life Conservation. In the arms control and nonproliferation area, Iran is a signatory to the Comprehensive Nuclear Test Ban Treaty, Biological Weapons Convention, Chemical Weapons Convention, International Atomic Energy Agency Safeguards Agreement, Partial Test Ban Treaty, and Treaty on the Non-Proliferation of Nuclear Weapons.

National Security

Armed Forces Overview

In 2007 the armed forces, under a unified command with the leaderas commander in chief, included about 420,000 active personnel in the regular forces and 125,000 in the auxiliary Islamic Revolutionary Guard Corps. Since the Iran–Iraq War of 1980–88, a main policy goal has been to be as independent as possible of alliances and foreign arms supplies. Accordingly, Iran has eschewed military alliances, although it has reached military supply agreements with a number of countries. Modernization of the navy, seen as vital for protecting interests in the Persian Gulf, is a high priority. Iran has purchased submarines and surface vessels from abroad and manufactures tanks, artillery, medium-range missiles, and helicopters. Technology purchased from North Korea and China, and refined by the domestic defense industry, supports a growing missile force that is considered the most important element of air defense policy.

Foreign Military Relations

In 2001 Iran signed a 10-year military-technical agreement with Russia that included assistance in aircraft maintenance and design estimated to be worth US$4 billion. In 2002 Iran signed a defense cooperation agreement with India. The latter agreement allows India to use Iranian military facilities in case of a war with Pakistan and provides Iran with Indian technical assistance in the building of aircraft and tanks. North Korea has supplied Iran with an unknown amount of technical assistance and equipment supporting the development of Iran's Shahab–3 ballistic missile.

External Threat

In 2003 the removal of Saddam Hussein from the presidency of Iraq eliminated a major regional threat, although during the 1990s United Nations sanctions on Iraq had limited the possibility of conflict. The events of 2003 replaced Saddam's threat with the large-scale, potentially long-term presence of the United States, a country that Iran officially considers its primary enemy. That event compounded the worry caused by the entry of U.S. forces into Afghanistan, to Iran's east, in 2001. Iran also considers Israel a major threat because Israel often has threatened an air strike against Iran's nuclear facilities.

Defense Budget

Iran's defense budget rose significantly from US$3.3 billion in 2004 and US$5.6 billion in 2005 to an estimated US$6.6 billion in 2006, or about 2.5 percent of gross domestic product.

Major Military Units

In 2007 the army had about 350,000 active personnel assigned to four armored divisions, six infantry divisions, two commando divisions, one airborne brigade, one special forces brigade, and six artillery groups. The navy had about 18,000 active personnel, including 2,600 in naval aviation and 2,600 in marine units. The navy operates bases at Bandar-e Abbas, Bushehr, Khark Island, Bandar-e Anzali, Bandar-e Khomeini, Bandar-e Mah Shahr, and Chabahar. The air force had about 52,000 active personnel, including 15,000 assigned to air defense units. Air force combat forces were organized in nine ground-attack fighter squadrons, five fighter squadrons, and one reconnaissance squadron.

Major Military Equipment

In 2006 the army had 1,613 main battle tanks, 610 armored infantry fighting vehicles, 640 armored personnel carriers, 2,010 pieces of towed artillery, 310 pieces of self-propelled artillery, 876 multiple rocket launchers, 5,000 mortars, 75 antitank guided weapons, 1,700 antiaircraft guns, and 50 attack helicopters. The navy had 3 submarines, 3 frigates, 140 patrol and coastal combatants, 5 mine warfare vessels, and 13 amphibious vessels. The air force ground-attack fighter units had F–4D, F–4E, F–5E, Su–24MK, Su–25K, and Mirage F–1E aircraft; the fighter units had F–14, F–7M, and MiG–29A aircraft. The air force also had 34 helicopters.

Military Service

Males 18 to 49 years of age are legally eligible for conscription, for an active service term of 18 months followed by voluntary reserve service. Individuals may volunteer for active duty at age 16. About 80 percent of army personnel are conscripts, as are 60 percent of the Revolutionary Guards; navy and air force personnel are mainly volunteers.

Paramilitary Forces

The volunteer paramilitary force, the Popular Mobilization Army, or Basij, has an estimated 300,000 personnel (including 40,000 in active service), mainly youths, with an estimated capability to expand to 1 million if needed. The Basij are under the authority of the Islamic Revolutionary Guard Corps.

Military Forces Abroad

In 2005 Iran had about 150 Revolutionary Guard Corps personnel in Lebanon, military advisers in Sudan, and three observers with the United Nations Mission in Ethiopia and Eritrea.

Police

About 40,000 police serve under the Ministry of Interior, including border patrol personnel. The Police–110 unit specializes in rapid-response activities in urban areas and dispersing gatherings deemed dangerous to public order. In 2003 some 400 women became the first female members of the police force since the 1978–79 Revolution.

Internal Threat

Despite strong government countermeasures, Iran is a main transit country for narcotics from neighboring Afghanistan and Pakistan and destined for Europe, Central Asia, and the Gulf region. Considerable quantities of these narcotics are sold illegally in Iran and are the main source of a serious and growing addiction problem. In the early 2000s, other types of smuggling increased rapidly, especially in Iran's impoverished border provinces. Corruption in the border police is a major factor in this trade. The Iraq-based Mojahedin-e Khalq (People's Fighters) seeks the removal of the Iranian regime by armed action, and its methods include the use of terrorist tactics. However, its activities inside Iran have been minimized by a domestic crackdown and by the fall of its patron, Saddam Hussein. Smaller insurgent groups are the Fedayan-e Khalq (People's Warriors), Democratic Party of Iranian Kurdistan, and Komela. The Paris-based National Council of Resistance of Iran is the umbrella group of the Mojahedin-e Khalq for overseas opposition groups.

Terrorism

During 2005 several incidents of domestic terrorism occurred using bombs planted in public places; most of these were in areas of ethnic tensions, such as West Azarbaijan (Kurds and Turks) and Khuzestan (Arabs and Lurs) provinces, although there also were bomb incidents in Tehran during the presidential election. Although Iran has consistently condemned all terrorist actions abroad, including those carried out against Israel, the U.S. Department of State named Iran the most active state sponsor of terrorism in the year 2004. Israel and the United States have contended that Iran has supplied funding, haven, training, and weapons for the Lebanese Hizballah and Palestinian groups such as Hamas and Palestinian Islamic Jihad.

Human Rights

International human rights organizations have cited major abuses in Iran's judicial system. Violations listed include arbitrary arrest, lack of due process, denial of access to attorneys, restrictions on family visits, prolonged periods in solitary confinement, and inhumane punishments in unofficial detention centers. Prison conditions are poor, particularly regarding food and medical care. The government has acknowledged its policy of incarcerating political enemies. The paramilitary Basij sometimes monitor behavior in public spaces and are authorized to detain individuals deemed in violation of dress codes.

After a period of liberalization in the first regime of President Khatami (1997–2001), in the early 2000s government control of the media grew noticeably more stringent. The government controls all television and radio broadcast facilities. Domestic and foreign publications and films are censored. The state also filters Internet content.

Members of religions not specifically protected by the constitution do not have full rights to assemble. Christianity, Judaism, and Zoroastrianism are the only protected minority religions. Lacking this designation, Iran's Bah'ais have been subjected to legal and religious discrimination and at times even persecution.

The Family Protection Law of 1967, which had extended some rights to women, was repealed after the 1978–79 Revolution. Marriage law discriminates against women in divorce, child custody, and inheritance from deceased spouses. In theory, a man may have as many as four wives at one time, provided he has the written consent of his wives; temporary marriages also are permitted, although they do not provide women with the same legal rights as permanent marriages. Court testimony by a woman is regarded as worth one-half that of a man. Although women have equal access to education, social and legal conditions limit their professional activities.

In: Iran Social, Economic and Political Developments
Editor: Joshua A. Owenstein
ISBN: 978-1-60741-552-7

Chapter 2

IRAN: U.S. CONCERNS AND POLICY RESPONSES*

Kenneth Katzman

SUMMARY

President Obama has said his Administration shares the goals of the previous Administration to contain Iran's strategic capabilities and regional influence, but the Obama Administration formulated approaches to achieve those goals that differ from those of its predecessor—in particular through expanded direct diplomatic engagement with Iran. This effort was put into practice with messages to the Iranian people by President Obama, and through invitations to and contact with Iranian diplomats at multilateral meetings, including those on Iran's nuclear program. The Administration also de-emphasized or discontinued policies that Iranian leaders considered hostile including: ratcheting up international sanctions, efforts to promote democracy in Iran, and openly discussing the potential for U.S. military action.

The Administration strategy on Iran is in some flux because of the allegations of a "stolen election" by the challengers to declared winner of the June 12, 2009 presidential election, incumbent President Mahmoud Ahmadinejad, and the crackdown against protesters who demanded a new vote. President Obama has criticized Iran's use of violence against protesters, and observers say that the Administration might try to take advantage of Iran's internal weakness to obtain a compromise that curbs Iran's nuclear program. If Iran refuses to return to the nuclear bargaining table, the Administration might focus, at an earlier than expected stage, on sanctioning and pressuring Iran. Bills in the 111th Congress, such as H.R. 2194 and S. 908, would tighten U.S. sanctions on Iran by amending the Iran Sanctions Act to penalize sales to Iran of gasoline and some of the measures contained in these bills have begun to advance as a consequence of the election-related violence. Congress also has passed resolutions that express solidarity with the demonstrators and condemn the regime's repression of them.

---

*This is an edited, reformatted and augmented version of a CRS Report for Congress publication dated July 2009.

The Obama Administration has not changed the previous Administration's characterization of Iran as a "profound threat to U.S. national security interests," a perception generated primarily by Iran's nuclear program and its military assistance to armed groups in Iraq and Afghanistan, to the Palestinian group Hamas, and to Lebanese Hezbollah. The U.S. policy approach was to try to prevent a nuclear breakout by Iran primarily by applying progressive multilateral economic pressure on Iran while also offering it potential cooperation should it suspend its enrichment of uranium. Multilateral efforts to pressure Iran include three U.N. Security Council resolutions (1737, 1747, and 1803) that ban weapons of mass destruction (WMD)-related trade with Iran; freeze the assets of Iran's nuclear entities and personalities; prevent Iran from transferring arms outside Iran; ban international travel by some Iranians; call for inspections of some Iranian sea and airborne cargo shipments; and call for restrictions on dealings with some Iranian banks. Separate U.S. efforts to persuade European governments to curb trade, investment, and credits to Iran; and to convince foreign banks not to do business with Iran, began to weaken Iran's economy, compounding the effect of lower oil prices since mid-2008. To emphasize its approach, the George W. Bush Administration built a substantial naval presence in the Persian Gulf.

Much of the debate over U.S. policy toward Iran has centered on the nature of the current regime; some believe that Iran, a country of about 70 million people, is a threat to U.S. Minterests because hardliners in Iran's regime dominate and set a policy direction intended to challenge U.S. influence and allies in the region. President George W. Bush, in his January 29, 2002, State of the Union message, labeled Iran part of an "axis of evil" along with Iraq and North Korea.

POLITICAL HISTORY

The United States was an ally of the late Shah of Iran, Mohammad Reza Pahlavi ("the Shah"), who ruled from 1941 until his ouster in February 1979. The Shah assumed the throne when Britain and Russia forced his father, Reza Shah Pahlavi (Reza Shah), from power because of his perceived alignment with Germany in World War II. Reza Shah had assumed power in 1921 when, as an officer in Iran's only military force, the Cossack Brigade (reflecting Russian influence in Iran in the early 20th century), he launched a coup against the government of the Qajar Dynasty. Reza Shah was proclaimed Shah in 1925, founding the Pahlavi dynasty. The Qajars had been in decline for many years before Reza Shah's takeover. That dynasty's perceived manipulation by Britain and Russia had been one of the causes of the 1906 constitutionalist movement, which forced the Qajars to form Iran's first Majles (parliament) in August 1906 and promulgate a constitution in December 1906. Prior to the Qajars, what is now Iran was the center of several Persian empires and dynasties, but whose reach shrunk steadily over time. Since the 16th century, Iranian empires lost control of Bahrain (1521), Baghdad (1638), the Caucasus (1828), western Afghanistan (1857), Baluchistan (1872), and what is now Turkmenistan (1894). Iran adopted Shiite Islam under the Safavid Dynasty (1500-1722), which brought Iran out from a series of Turkic and Mongol conquests.

The Shah was anti-Communist, and the United States viewed his government as a bulwark against the expansion of Soviet influence in the Persian Gulf and a counterweight to pro-Soviet Arab regimes and movements. Israel maintained a representative office in Iran during the Shah's time and the Shah supported a peaceful resolution of the Arab-Israeli dispute. In 1951, under pressure from nationalists in the Majles (parliament) who gained strength in the 1949 Majles elections, he appointed a popular nationalist parliamentarian, Dr. Mohammad Mossadeq, as Prime Minister. Mossadeq was widely considered left-leaning, and the United States was wary of his policies, which included his drive for nationalization of the oil industry. Mossadeq's followers began an uprising in August 1953 when the Shah tried to dismiss Mossadeq, and the Shah fled. The Shah was restored in a successful CIA-supported uprising against Mossadeq.

The Shah tried to modernize Iran and orient it toward the West, but in so doing he also sought to marginalize Iran's Shiite clergy. He exiled Ayatollah Ruhollah Khomeini in 1964 because of Khomeini's active opposition, which was based on the Shah's anti-clerical policies and what Khomeini alleged was the Shah's forfeiture of Iran's sovereignty to the United States. Khomeini fled to and taught in Najaf, Iraq, a major Shiite theological center that contains the Shrine of Imam Ali, Shiism's foremost figure. There, he was a peer of senior Iraqi Shiite clerics and, with them, advocated direct clerical rule or *velayat-e-faqih* (rule by a supreme Islamic jurisprudent). In 1978, three years after the March 6, 1975, Algiers Accords between the Shah and Iraq's Baathist leaders, which settled territorial disputes and required each party to stop assisting each other's oppositionists, Iraq expelled Khomeini to France, from which he stoked the Islamic revolution. Mass demonstrations and guerrilla activity by pro-Khomeini forces, allied with a broad array of anti-Shah activists, caused the Shah's government to collapse in February 1979. Khomeini returned from France and, on February 11, 1979, declared an Islamic Republic of Iran, as enshrined in the constitution that was adopted in a public referendum in December 1979 (and amended in 1989). Khomeini was strongly anti-West and particularly anti-U.S., and relations between the United States and the Islamic Republic turned hostile even before the November 4, 1979, seizure of the U.S. Embassy by pro-Khomeini radicals.

Regime Structure, Stability, and Elections

About a decade after founding the Islamic republic, Ayatollah Ruhollah Khomeini died on June 3, 1989. The regime he established—enshrined in an Islamic republican constitution adopted in 1980 and amended in 1989—consists of some elected and some appointed positions. It has appeared relatively stable, but, even before the June 12, 2009, presidential election, faced periodic but low-level unrest in areas inhabited by minorities, and substantial unpopularity among many intellectuals, students, some labor groups, and women. Although Iran's is widely considered to be an authoritarian regime, the various centers of power are able, politically and constitutionally, to offer a degree of checks and balances and popular input. National elections under the Islamic republic have always been held, and on time, even during the eight year Iran-Iraq war, although there are limitations on who can run and the election process is run by the government.

The Supreme Leader, His Powers, and Other Ruling Councils

Upon Khomeini's death, one of his disciples, Ayatollah Ali Khamene'i, was selected Supreme Leader by an elected 86-seat "*Assembly of Experts*."[1] The Assembly of Experts is empowered to oversee the work of the Supreme Leader and replace him if necessary, as well as to amend the constitution. The Assembly serves a six-year term; the fourth election for that Assembly was held on December 15, 2006. After that election, Akbar Hashemi-Rafsanjani, still a major figure having served two terms as president himself (1989-1997), was named deputy leader of the Assembly. After the death of the leader of the Assembly, Rafsanjani was selected its head in September 2007, outpointing a harder line competitor, Ayatollah Ahmad Jannati. Khamene'i has vast formal powers as Supreme Leader—he is Commander in Chief of the armed forces, giving him the power to appoint commanders and to be represented on the highest national security body, the Supreme National Security Council, composed of top military and civilian security officials. He appoints half of the twelve-member *Council of Guardians*;[2] and the members of Iran's Supreme Judicial Council. Headed by Jannati, the conservative-controlled Council of Guardians reviews legislation to ensure it conforms to Islamic law, and it screens election candidates and certifies elections results. The Supreme Leader also has the power, under the constitution, to remove the elected President if either the Supreme Judicial Council or the elected *Majles* (parliament) say the President should be removed, with cause.

The Supreme Leader appoints members of the 42-member *Expediency Council*, set up in 1988 to resolve legislative disagreements between the *Majles* and the Council of Guardians but its powers were expanded in 2006 to include oversight of the executive branch (cabinet) performance. Expediency Council members serve five-year terms. The Council, appointed most recently in February 2007, is still headed by Rafsanjani; its executive officer is former Revolutionary Guard commander-in-chief Mohsen Reza'i.

Table 1. Major Factions and Personalities

Conservatives	
Supreme Leader Ali Khamene'i	Born in July 1939 to an Azeri (Turkic) family from Mashhad. Lost the use of his right arm in an assassination attempt in June 1981. Helped organize the Revolutionary Guard and other post-revolution security organs. Served as elected president during 1981-1989 and was selected Khomeini's successor in June 1989 upon the Ayatollah's death. Upon that selection, his religious ranking was advanced in the state-run press and official organs to "Ayatollah" from the lower ranking "Hojjat ol-Islam." Has all the formal powers but not the undisputed authority of his predecessor, founder of the revolutionary regime Ayatollah Ruhollah Khomeini. Like Khomeini, Khamene'i generally stays out of day-to-day governmental business but saves his prestige to resolve factional disputes or to quiet popular criticism of regime performance. Considered moderate-conservative on domestic policy but hardline on foreign policy and toward Israel. Seeks to challenge U.S. hegemony and wants Israel defeated but respects U.S. military power and fears military confrontation with United States. Generally supports the business community (bazaaris), and opposes state control of the economy.

Conservatives	
Ali Akbar Hashemi-Rafsanjani	Long a key strategist of the regime, and longtime advocate of "grand bargain" to resolve all outstanding issues with United States, although on Iran's terms. A mid-ranking cleric, now leads both Expediency Council and Assembly of Experts, although generally perceived as waning in influence generally. Heads moderate-conservative faction known as Executives of Construction. Was Majles speaker during 1981-89 and President 1989-1997. One of Iran's richest men, family owns large share of Iran's total pistachio nut production. Supported Musavi in June 2009 election, purportedly financed much of his campaign. Playing behind-the-scenes role trying to persuade Supreme Leader to nullify the June 2009 election, but arrest of five family members on June 20, 2009, reflects Supreme Leader's pressure on him.
President Mahmoud Ahmadinejad	Re-elected on June 12, 2009, but results disputed by his election challengers. See box .
Majles Speaker Ali Larijani	Overwhelming winner for Majles seat from Qom on March 14, 2008, and selected Majles Speaker on May 25 (237 out of 290 votes). Former state broadcasting head (1994-2004) and Minister of Culture and Islamic Guidance (1993) , was head of Supreme National Security Council and chief nuclear negotiator from August 2005 until Otober 2007 resignation. Sought to avoid U.N. Security Council isolation. Politically close to Khamene'i but has criticized election officials for the flawed June 12, 2009, election.
Mohammad Baqer Qalibaf	Former Revolutionary Guard Air Force commander and police chief, but a moderate- conservative. Encourages comparisons of himself to Reza Shah, invoking an era of stability and strong leadership, while also making use of modern media tools. Lost in the 2005 presidential elections, but supporters won nine out of 15 seats on Tehran city council in December 2006 elections, propelling him to current post as mayor of Tehran. Recruited moderate conservatives for March 2008 Majles election.
Ayatollah Mohammad Taqi Mesbah-Yazdi	Founder of the hardline Haqqani school, and spiritual mentor of Ahmadinejad. Fared poorly in December 2006 elections for Assembly of Experts. An assertive defender of the powers of the Supreme Leader and a proponent of an "Islamic state" rather than the current "Islamic republic," and advocates isolation from the West. May seek to replace Khamene'i.
Ayatollah Mahmud Hashemi Shahrudi	Has headed the Supreme Judicial Council since 1999. Ally of Khamene'i and Rafsanjani, has supported repeated crackdowns on independent media critical of the regime. But, has cracked down on judicial corruption and on mistreatment of prisoners. Politically close to Shiite Islamist parties in Iraq, he is hosting Moqtada Al Sadr, the radical young Iraqi cleric who has been studying Islamic theology in Qom since late 2007.
Militant Clerics Association	Longtime organization of hardline clerics headed by Ayatollah Mohammad Mahdavi-Kani. Not to be confused with an organization with almost the same name, below. Did not back Ahmadinejad in June 12 presidential elections.

Table 1. (Continued)

Reformists	
Mohammad Khatemi/Mir Hossein Musavi	Khatemi - reformist president during 1997-2005 and declared he would run again for President in June 2009 elections, but withdrew when allied reformist Mir Hossein Musavi entered the race in late March 2009. Khatemi elected May 1997, with 69% of the vote; re-elected June 2001with 77%. Rode wave of sentiment for easing social and political restrictions among students, intellectuals, youths, and women that seeks reform but not outright replacement of the regime, but became disillusioned with Khatemi failure to stand up to hardliners on reform issues. Now heads International Center for Dialogue Among Civilizations. Visited U.S. in September 2006 to speak at Harvard and the Washington National Cathedral on "dialogue of civilizations." Has hewed to staunch anti- Israel line of most Iranian officials, but perceived as open to accepting a Palestinian-Israeli compromise. Musavi has views similar to Khatemi on political and social freedoms and on reducing Iran's international isolation, but supports strong state intervention in the economy to benefit workers, lower classes. Khatemi supports Musavi challenge to 2009 election legitimacy.
Society of Militant Clerics	Reformist grouping once led by Mehdi Karrubi. Karrubi formed a separate "National Trust" faction after losing 2005 election. Ran again in 2009, but received few votes and is challenging election legitimacy. Society has backed some election-related protests.
Office of Consolidation Unity (Daftar Tahkim-e-Vahdat)	Staunch reformists. Originally strong Khatemi supporters, but turned against him for failing to challenge hardliners, particularly after July 1999 violent crackdown on student riots, in which four students were killed. Generally dispersed and repressed under Ahmadinejad. Generally supported Karrubi in 2009 elections and supports protests.
Islamic Iran Participation Front (IIPF).	The most prominent and best organized pro-reform grouping. Its leaders include Khatemi's brother, Mohammad Reza Khatemi (a deputy speaker in the 2000-2004 Majles) and Mohsen Mirdamadi. Backed Musavi in June 12 election; several IIPF leaders detained in post-election dispute.
Mojahedin of the Islamic Revolution Organization (MIR)	Composed mainly of left-leaning Iranian figures who support state control of the economy, but want greater political pluralism and relaxation of rules on social behavior. A major constituency of the reformist camp. Its leader is former Heavy Industries Minister Behzad Nabavi, who supported Musavi in 2009 election and purportedly was detained.

The Presidency/Mahmoud Ahmadinejad

The President, a position held since 2005 by Mahmoud Ahmadinejad, appoints the cabinet, but nominees are subject to confirmation by the Majles (parliament). Although subordinate to the Supreme Leader, the presidency is a coveted and intensely fought-over position which provides vast opportunities for the President to empower his political base and to affect policy.

After suffering several election defeats at the hands of President Mohammad Khatemi and the reformists in the 1997 and 2001 presidential elections, hardliners successfully moved to regain the sway they held when Khomeini was alive. Conservatives won the February 20, 2004, Majles elections (which are always held one year prior to each presidential election), although the conservative win was the result of the Council of Guardians' disqualification of 3,600 reformist candidates, including 87 Majles incumbents. That helped conservatives win 155 out of the 290 seats. The George W. Bush Administration and the Senate (S.Res. 304, adopted by unanimous consent on February 12, 2004) criticized the elections as unfair.

As the reformist faction suffered setbacks even with Khatemi in the presidency, the Council of Guardians narrowed the field of candidates for the June 2005 presidential elections to 8 out of the 1,014 persons who filed. Rafsanjani [3] was considered the favorite against several opponents more hardline than he is—three had ties to the Revolutionary Guard: Ali Larijani (see Table 1); Mohammad Baqer Qalibaf (see Table 1); and Tehran mayor Mahmoud Ahmadinejad. In the June 17, 2005 first round, turnout was about 63% (29.4 million votes out of 46.7 million eligible voters). With 21% and 19.5%, respectively, Rafsanjani and Ahmadinejad, who did unexpectedly well because of tacit backing from Khamene'i, moved to a run-off. Reformist candidates Mehdi Karrubi and Mostafa Moin fared worse than expected. Ahmadinejad won in the June 24 runoff, receiving 61.8% to Rafsanjani's 35.7%. He took office on August 6, 2005.

Ahmadinejad's Policies and Popularity

Ahmadinejad has been a controversial figure for statements often perceived as inflammatory. He notably attracted opprobrium at an October 26, 2005, Tehran conference entitled "A World Without Zionism" by stating that "Israel should be wiped off the map." Another point of contention was his insistence on the holding of a December 2006 conference in Tehran questioning the Holocaust, a theme he has returned to several times since, including at a September 2007 speech at Columbia University. A U.N. Security Council statement and Senate and House resolutions (H.Res. 523 and S.Res. 292), passed by their respective chambers, condemned the statement. On June 21, 2007, the House passed H.Con.Res. 21, calling on the United Nations Security Council to charge Ahmadinejad with violating the 1948 Convention on the Prevention and Punishment of the Crime of Genocide; the Convention includes "direct and public incitement" to commit genocide as a punishable offense.

Even before the 2009 presidential election campaign, several Iranian leaders, and portions of the population, were expressing concern that Ahmadinejad's defiance of the international community on the nuclear issue—as well as his frequent visits and meetings with such anti-U.S. figures as Venezuela's Hugo Chavez — was isolating Iran. The perceived failings of Ahmadinejad caused his conservative faction to split in the March 2008 Majles elections. On November 4, 2008, the Majles impeached Interior Minister Ali Kordan for falsely claiming he had an Oxford University degree. Sadeq Mahsouli, a former Revolutionary Guard official, was confirmed as replacement on November 18, 2008, and his office oversaw the June 12, 2009, presidential election.

Throughout Ahmadinejad's first term, Supreme Leader Khamene'i issued public statements of support for Ahmadinejad to rally the regime against international pressure. In August 2008, he praised Ahmadinejad for refusing to bow to international demands on the nuclear issue and said the cabinet should make plans for another four years. At other times, such as April 2009, Khamene'i has upbraided Ahmadinejad – in this case for incorporating the position of coordinator of the Hajj (major pilgrimage to Mecca) into the Tourism

Ministry; the move was reversed in response to Khamene'i's criticism. Khamene'i was perceived as favoring Ahmadinejad's reelection but, sensing that this outcome was not assured, he was publicly neutral in the campaign.

On economic matters, many Iranian figures criticized Ahmadinejad for raising some wages and lowering interest rates for poorer borrowers, cancelling some debts of farmers, and increasing social welfare payments and subsidies. These moves fed inflation, but poorer Iranians saw Ahmadinejad as attentive to their economic plight and this support appears to have been key to his re-election, if the results hold. Iranian economists say that these social welfare programs have caused a diminishment of Iran's reserve fund ("Oil Stabilization Fund") even when oil prices were high in mid-2008, leaving Iran unable to cope with the fall in oil prices. Others say he has not moved to curb the dependence on oil revenues, which account for about 20% of Iran's gross domestic product (GDP), and its extensive imports of refined gasoline. Major economic sectors or markets are controlled by the quasi-statal "foundations" (*bonyads*), run by powerful former officials, and there are special trading privileges for them and the bazaar merchants, a key constituency for some conservatives. The same privileges reportedly apply to businesses run by the Revolutionary Guard, as discussed below.

Ahmadinejad has generally been opposed by affluent and educated urbanites. Urban sentiment against him was belied in protests in November 2008 by bazaar merchants in major cities against the imposition of sales taxation; the protests caused a suspension of the imposition of the tax. Some business owners say the difficulty obtaining credit from foreign banks is hurting their ability to operate. Students have conducted several high-profile anti-Ahmadinejad protests in recent years, most recently in late February 2009 when authorities tried to rebury on Amir Kabir University of Technology grounds the bodies of some killed in the Iran-Iraq war.

MAHMOUD AHMADINEJAD

First non-cleric to be president of the Islamic republic since the assassination of then president Mohammad Ali Rajai in August 1981. About 55, he asserts he is a "man of the people," the son of a blacksmith who lives in modest circumstances, who would promote the interests of the poor and return government to the original principles of the Islamic revolution. Has burnished that image as president through regular visits to poor areas and through subsidies directed at the lower classes. His official biography says he served with the "special forces" of the Revolutionary Guard, and he served subsequently (late 1980s) as a deputy provincial governor. A part of the "Isargaran" faction composed of former Guard and Basij (volunteer popular forces) leaders and other hardliners. U.S. intelligence reportedly determined he was not one of the holders of the 52 American hostages during November 1979-January 1981. Other accounts say Ahmadinejad believes his mission is to prepare for the return of the 12th Imam—Imam Mahdi—whose return from occultation would, according to Twelver Shiite doctrine, be accompanied by the establishment of Islam as the global religion. Earned clerical criticism in May 2008 for again invoking intervention by Imam Mahdi in present day state affairs. Regularly attends U.N. General Assembly sessions in New York each September. In an October 2006 address, Ahmadinejad said, "I have a connection with God." Sent letter of congratulation to

President-elect Barack Obama for his election victory, but has only tepidly responded to subsequent Obama Administration outreach initiatives. Following limited recount, declared winner of June 12, 2009, election.

June 12, 2009, Presidential Elections

The reformists' prospects in the election appeared to brighten in February 2009, when Khatemi—who feared that his running again would begin a potentially wrenching and divisive political battle between conservatives and reformists — said that he would run. However, on March 18, 2009, Khatemi withdrew from the race when another major reformist, former Prime Minister Mir Hossein Musavi, said he would run. Khatemi backed Musavi enthusiastically.

A total of about 500 candidates for the June 12, 2009, presidential elections registered their names during May 5-10, 2009. The Council of Guardians decide on the final candidates on May 20—permitting only four to run: Ahmadinejad, Mir Hossein Musavi, Mehdi Karrubi, and Mohsen Reza'i. This was, to some extent, different from past elections in which the field consisted of about 8-10 candidates. Forty women registered, but, as in the past, the Council interpreted the constitution as prohibiting women from running for president. They can run for the Majles and other elected jobs, however. The Interior Ministry, which runs the election, also instituted during this campaign season a series of one-on-one debates among the candidates, which were acrimonious, including Ahmadinejad's accusations of corruption against Rafsanjani and against Musavi's wife. If no candidate received more than 50% of the vote on June 12, there would have been a runoff on June 19.

Aside from Ahmadinejad, the candidates were as follows:

Mir Hosein Musavi. The main reformist candidate. Non-cleric. About 67. Architect and disciple and aide of Ayatollah Khomeini, he served as Foreign Minister (1980), then Prime Minister (1981-89), at which time he successfully managed the state rationing program during the privations of the Iran-Iraq war but often feuded with Khamene'i, who was then President. At that time, he was an advocate of state control of the economy. His post was abolished in the 1989 revision of the constitution. Later moderated his views, and shares much of Khatemi's policy outlook, including the need to avoid confrontation with the international community. However, Musavi strongly opposed U.N.-demanded curbs on Iran's nuclear program. Musavi was believed to benefit by having been out of politics since he left the prime ministership in 1989 and was therefore untainted by recent allegations of corruption or crackdown on civil society groups. Musavi's campaign made extensive use of his high profile wife, Zahra Rahnevard, a well-known women's activist and professor. He continues to contest the election results, but now says any further protests should have official permission.

Table 2. Factions in the Eighth Majles (Elected March 14-April 25, 2008)

Pro-Ahmadinejad Conservatives (United Front of Principalists)	117
Anti-Ahmadinejad Conservatives (Coalition of Principalists)	53
Reformists (39 seats in eighth Majles)	46
Independents	71
Seats annulled or voided	3

Mehdi Karrubi. Some feared he might split the reformist vote because of his attentiveness to economic policies that favor the lower classes, but official results showed him a minor factor in the voting. He continues to question the legitimacy of the election.

Mohsen Reza'i. As noted above, he was Commander in Chief of the Revolutionary Guard for almost all of the Iran-Iraq war period. About 57 years old, he is considered an anti-Ahmadinejad conservative. Reza'i dropped out just prior to the 2005 presidential election due to perceived insufficient support, and he apparently did not build substantial support since then. He attended Khamene'i's June 19 speech and later dropped his formal challenge of the election results.

Election Dispute and Aftermath

The outcome of the election was always difficult to foresee. Polling results were inconsistent. Musavi supporters held large rallies in Tehran and elsewhere, suggesting momentum, although pro-Ahmadinejad rallies were large as well. During the campaign, Khamene'i met with Musavi and, in mid-May 2009, visited Musavi's father at his home, suggesting neutrality. Both Khamene'i and Musavi are from the same home town (Khamein), although the two were often at odds during the Iran-Iraq war, when Khamene'i was President and Musavi was Prime Minister. t Ahmadinejad supporters viewed him as standing firm against U.S. pressure, and believed Musavi would bow to international pressure on the nuclear issue and on foreign policy.

The vote turnout was unexpectedly high at about 85%: 39.1 million valid (and invalid) votes were cast. The Interior Ministry announced two hours after the polls closed that Ahmadinejad had won, although in the past results have been announced the day after. The totals were announced on Saturday, June 13, 2009, as follows:

Ahmadinejad: 24.5 million votes – 62.6%
Musavi: 13.2 million votes – 33.75%
Reza'i: 678,000 votes – 1.73%
Invalid: 409,000 votes – 1%
Karrubi: 333,600 votes – 0.85%

Almost immediately after the results were announced, Musavi supporters began protesting the results on June 13, as he, Karrubi, and Reza'i, asserted outright fraud and called for a new election, citing the premature release of results; the barring of candidate observers at some polling stations; regime shut-down of internet and text services; and repression of post-election protests. Khamene'i declared the results a "divine assessment," appearing to certify the results even though formal procedures require a three day complaint period. While several outside analysts say the results appeared to represent widespread fraud.[4], others said the announced results were realistic and reflected Ahmadinejad's perceived strong support in rural areas and among the urban poor. Those who take this view note that Musavi did win in Tehran city where his popularity was purportedly centered, according to the Interior Ministry's province-by-province results released June 16.

Protests built throughout June 13-19, although far larger in Tehran than in other cities, and regime security forces used varying amounts of force to control them, causing 17 deaths, according to official Iranian statements. However, the protest movement's hopes of having Khamene'i annul the election were dashed by his major Friday prayer sermon on June 19 in which he refuted allegations of widespread fraud and implicitly threatened a crackdown on any further protests. That crackdown was in evidence against protesters on Saturday, June 20, with state media reporting at least ten killed that day. However, with Musavi calling for continued demonstrations but with exercise of restraint, the protests appeared to wane by June 22. The regime's attempts to black out international media and internet access to Iran continued, with mixed success, and it has arrested a reputed 1,000 reformist leaders, although claiming to have released most by now.

In order to quiet protests, the regime tried to at least appear to address the complaints. The Council of Guardians agreed to the review the results in 650 polling stations, and, on June 22, it said that in 50 districts, there were more votes counted than the number of voters but that this irregularity did not appear to affect the outcome of the election. Later, on June 29, 2009, the Council recounted 10% of the votes of Tehran's districts and some provincial ballots as well, in the presence of state television, and, finding no irregularities, certified the results. Musavi and Karrubi, on websites, continued to call the election fraudulent and did not send observers to the partial recount.

Some say that the most serious effects were the widening of cracks within the regime, creating the most serious internal rift in Iran since the early 1980s. Rafsanjani backed the Musavi challenge, reportedly in part by appealing—to date unsuccessfully—to senior clerics to try to pressure Khamene'i to annul the vote. His activities, as well as those of his family members, who were supporting the Musavi effort and the protests, in part caused several Rafsanjani children to be arrested and detained on June 20. 2009. He later appeared to back down by signing on to an Expediency Council statement on June 27, 2009 urging that challenges to the election pursue legal processes. Larijani, Qalibaf, and several senior Ayatollahs criticized the use of violence against the protesters. An open question is whether Ahmadinejad might try to heal the internal rift when he appoints a new cabinet following his July/August second inauguration.

U.S. and Allied Reaction

The dispute caused a dilemma for President Obama who has tried to balance non-interference in Iranian affairs (a sensitive issue in Iran) with the need to stress U.S. values. He steered clear of alleging wholesale fraud but, at least initially, indicated that U.S. policy would not change and that Musavi held policy positions little different than those of Ahmadinejad. Possibly reflecting criticism that the Administration did not strongly side with the protests, and that U.S. reaction was milder than that of most European governments, his statements became somewhat more critical, including a June 23 comment that condemned the violence and a June 26 comment that Administration diplomacy would be affected by the Iranian crackdown. The increased U.S. criticism caused Iran to accelerate its charges that the West was meddling in Iran's turmoil and to demand a U.S. apology for the criticism. The later Obama statements appeared to have been influenced, to some extent by House and Senate passage of resolutions on June 19 (H.Res. 560 and S.Res. 193, respectively), condemning violence against demonstrators and the government's suppression of independent electronic communication. Another resolution passed by the Senate that day, S.Res. 196, calling on the Iranian regime to permit freedom of expression, freedom of speech, and freedom of the press. More information on how the dispute might affect Obama Administration policy is addressed in the section "Implementation of the Engagement Policy," below.

Table 3. Selected Economic Indicators

Economic Growth	6.4% (2008 est.)
Per Capita Income	$13,100/yr purchasing power parity
Proven Oil Reserves	135 billion barrels (highest after Russia and Canada)
Oil Production/Exports	4.0 million barrels per day (mbd)/ 2.4 mbd exports. Exports could shrink to zero by 2015-2020 due to accelerating domestic consumption.
Major Oil/Gas Customers	China—300,00 barrels per day (bpd); about 4% of China's oil imports; Japan—600,000 bpd, about 12% of oil imports; other Asia (mainly South Korea)—450,000 bpd; Italy— 300,000 bpd; France—210,000 bpd; Netherlands 40,000 bpd; other Europe—200,000 bpd; India—150,000 bpd (10% of its oil imports; Africa—200,000 bpd. Turkey—gas: 8.6 billion cubic meters/yr
Refined Gasoline Import/ Suppliers	Imports were $5 billion value per year in 2006, but now about $4 billion per year after rationing. Traders and suppliers include: Vitol (Switzerland), which supplies about 30% of Iran's gasoline; Total (France); Trafigura (Switzerland/Nethelands); Reliance Energy (India, Jamnagar refinery); Russia's Lukoil; Kuwait, UAE, Turkey, Venezuela (Petroleos de Venezuela), Singapore, the Netherlands, China, Turkmenistan, and Azerbaijan. Iran planning at least eight new or upgrade refinery projects to expand capacity to about 3 million barrels per day from 1.5 mbd.
Major Export Markets (2006)	Japan ($9.9 billion); China ($9.2 billion); Turkey ($5.1 billion); Italy ($4.45 billion); South Korea ($4 billion); Netherlands ($3.2 billion); France ($2.7 billion); South Africa ($2.7 billion); Spain ($2.3 billion); Greece ($2 billion)
Major Imports (2006)	Germany ($5.6 billion); China ($5 billion); UAE ($4 billion); S. Korea ($2.9 billion); France ($2.6 billion); Italy ($2.5 billion); Russia ($1.7 billion); India ($1.6 billion); Brazil ($1.3 billion); Japan ($1.3 billion).
Export Credit Guarantees (2006)	Germany $715 million, down from $2 billion in 2005; France—$3.8 billion, down from $5.7 billion in 2005.
Major Non-Oil Investments	Renault (France) and Mercedes (Germany)-automobile production in Karaj, Iran—valued at $370 million; Renault (France), Peugeot (France) and Volkswagen (Germany)—auto parts production; Turkey—Tehran airport, hotels; China—shipbuilding on Qeshm Island, aluminum factory in Shirvan, cement plant in Hamadan; UAE financing Esfahan Steel Company; India—steel plant, petrochemical plant; S. Korea—steel plant in Kerman Province; S. Korea and Germany—$1.7 billion to expand Esfahan refinery.
Trade With U.S. (2008)	$785 million (trade is severely restricted by U.S. sanctions). Exports to U.S.—$102 million (large categories: pomegranate juice, caviar, pistachio nuts, carpets, medicines, artwork). Imports from U.S.—$683 million (wheat: - $535 million; medicines, tobacco products, seeds).
"Oil Stabilization Fund" Reserves	$12.1 billion (August 2008, IMF estimate).
External Debt	$19 billion (2007 est.)
Development Assistance Received	2003 (latest available): $136 million grant aid. Biggest donors: Germany ($38 million); Japan ($17 million); France ($9 million).
Inflation	15% + (May 2009), according to Iranian officials.
Unemployment Rate	11%+

Source: CIA World Factbook, various press, IMF, Iran Trade Planning Division, press.

As noted, several European governments, such as France, Britain, and Germany, have been highly critical of Iran's crackdown on protests. As of July 2, the European Union is considering recalling its ambassadors from Iran in solidarity with Britain. In late June, Iran arrested eight Iranian employees of the British Embassy in Tehran on charges of fomenting the protests; all but one have been released as of July 1.

HUMAN RIGHTS AND DISSENT

Prior to the election unrest, the regime appeared to have a firm grip on power, in part because it vigorously suppresses dissent. Some observers believe that the election-related unrest could still revive and pose a threat to the regime. Most experts believed that a victory by Musavi in the June 12 election would have partly satisfied the demands of these groups for more freedoms.

The latest State Department human rights report (released February 25, 2009), the 2008 State Department "religious freedom" report (released September 19, 2008), and a report by the U.N. Secretary General on October 1, 2008, cite Iran for widespread serious abuses, including unjust executions, politically motivated abductions by security forces, torture, arbitrary arrest and detention, and arrests of women's rights activists. The State Department human rights reports said the government's "poor human rights record worsened" during 2008. The Secretary General's report later became the basis of a U.N. General Assembly resolution, finalized on December 18, 2008, by a vote of 69-54, calling on Iran to allow visits by U.N. personnel investigating the status of human rights practices in Iran. Subsequent to the passage of the resolution, Iranian authorities raided the Tehran office of the Center for Defenders of Human Rights, headed by Nobel Peace Prize laureate (2003) and Iran human rights activist lawyer Shirin Abadi, who reportedly has left Iran fearing arrest in connection with the post-election dispute.

The regime has been particularly concerned about dissidents who previously held senior regime positions; these dissidents have followings inside Iran. One figure, Ayatollah Hossein Ali Montazeri, was released in January 2003 from several years of house arrest, but he remains under close watch. Montazeri was Khomeini's designated successor until 1989, when Khomeini dismissed him for allegedly protecting intellectuals and opponents of clerical rule.

Other dissidents have sought to challenge or expose the regime's practices from inside Iran, mainly focused on human rights and free speech. Journalist Akbar Ganji conducted hunger strikes to protest regime oppression; he was released on schedule on March 18, 2006 after sentencing in 2001 to six years in prison for alleging high-level involvement in 1999 murders of Iranian dissident intellectuals that the regime had blamed on "rogue" security agents.

Table 4. Human Rights Practices

Group/Issue	Regime Practice/Recent Developments
Ethnic and Religious Breakdown	Persians are about 51% of the population, and Azeris (a Turkic people) are about 24%. Kurds are about 7% of the population, and about 3% are Arab. Of religions, Shiite Muslims are about 90% of the Muslim population and Sunnis are about 10%. About 2% of the population is non-Muslim, including Christians, Zoroastrians (an ancient religion in what is now Iran), Jewish, and Baha'i.
Media	Since 2000, judicial hardliners have closed hundreds of reformist newspapers, although many have tended to reopen under new names. Iran also has blocked hundreds of pro-reform websites, and hardliners reportedly are blocking websites and blogs supportive of the presidential candidacy of Khatemi. In August 2007, the government closed a major reformist daily newspaper, *Shargh*, which had previously been suspended repeatedly. In February 2008, the regime closed the main women's magazine, Zanan (women in Farsi) for allegedly highlighting gender inequality in Islamic law. In November 2008, the regime arrested famed Iranian blogger Hossein Derakshan. Canadian journalist (of Iranian origin) Zahra Kazemi was detained in 2003 for filming outside Tehran's Evin prison and allegedly beaten to death in custody. The intelligence agent who conducted the interrogation/beating was acquitted July 25, 2004.
Labor Unions/Students/ Other Activists	Independent unions are technically legal but not allowed in practice. The sole authorized national labor organization is a state-controlled "Workers' House" umbrella. However, some activists show independence and, in 2007, the regime arrested labor activists for teachers' associations, bus drivers' unions, and a bakery workers' union. A bus drivers union leader, Mansur Osanloo, has been in jail since July 2007. The regime reportedly also dissolved student unions and replaced them with regime loyalists following student criticism of Ahmadinejad. In September 2008, Iran arrested several HIV/AIDs researchers for alleged anti-government activities.
Women	Regime strictly enforcing requirement that women fully cover themselves in public, generally with a garment called a *chador,* including through detentions. In March 2007, the regime arrested 31 women activists who were protesting the arrest in 2006 of several other women's rights activists; all but 3 of the 31 were released by March 9. In May 2006, the Majles passed a bill calling for increased public awareness of Islamic dress, an apparent attempt to persuade women not to wear Western fashion. The bill did not contain a requirement that members of Iran's minority groups wear badges or distinctive clothing. In April 2006, Ahmadinejad directed that women be allowed to attend soccer matches, but the Supreme Leader reversed that move. Women can vote and run in parliamentary and municipal elections. Iranian women can drive, and many work outside the home, including owning their own businesses. There are 9 women in the 290-seat Majles.
Religious Freedom	Each year since 1999, the State Department religious freedom report has named Iran as a "Country of Particular Concern" under the

Table 4. (Continued)

Group/Issue	Regime Practice/Recent Developments
	International Religious Freedom Act. No sanctions added, on the grounds that Iran is already subject to extensive U.S. sanctions. Continued deterioration in religious freedom noted in the International Religious Freedom report for 2008 (September 19, 2008).
Baha'is	Iran repeatedly cited for repression of the Baha'i community, which Iran's Shiite Muslim clergy views as a heretical sect. The State Department cited Iran on February 13, 2009, for charging seven Bahai's with espionage; thirty other Bahai's remain imprisoned. In the 1990s, several Baha'is were executed for apostasy (Bahman Samandari in 1992; Musa Talibi in 1996; and Ruhollah Ruhani in 1998). Another, Dhabihullah Mahrami, was in custody since 1995 and died of unknown causes in prison in December 2005. A wave of Baha'i arrests occurred in May 2006 and two-thirds of university students of the Baha'i faith were expelled from university in 2007. Several congressional resolutions have condemned Iran's treatment of the Baha'is, including in 1982, 1984, 1988, 1990, 1992, 1994, 1996, 2000, and 2006. In the 110th Congress, H.Res. 1008 condemned Iran's treatment of the Baha'is (passed House August 1, 2008).
Jews	Along with Christians, a "recognized minority," with one seat in the Majles, the 30,000-member Jewish community (the largest in the Middle East aside from Israel) enjoys somewhat more freedoms than Jewish communities in several other Muslim states. However, in practice the freedom of Iranian Jews to practice their religion is limited, and Iranian Jews remain reluctant to speak out for fear of reprisals. During 1993-1998, Iran executed five Jews allegedly spying for Israel. In June 1999, Iran arrested 13 Jews (mostly teachers, shopkeepers, and butchers) from the Shiraz area that it said were part of an "espionage ring" for Israel. After an April-June 2000 trial, ten of the Jews and two Muslims accomplices were convicted (July 1, 2000), receiving sentences ranging from 4 to 13 years. An appeals panel reduced the sentences, and all were released by April 2003. On November 17, 2008, Iran hanged businessman Ali Ashtari (a Muslim), who was arrested in 2006, for allegedly providing information on Iran's nuclear program to Israel.
Sunnis	The cited reports note other discrimination against Sufis and Sunni Muslims, although abuses against Sunnis could reflect that minority ethnicities, including Kurds, are mostly Sunnis.
Human Trafficking	The June 16, 2009, (latest annual), State Department "Trafficking in Persons" report continues to place Iran in Tier 3 (worst level) for failing to take action to prevent trafficking in persons. Girls are trafficked for sexual exploitation within Iran and from Iran to neighboring countries.
Juvenile Executions	Iran executed six persons under the age of 18 in 2008, the only country to do so. As a party to the International Covenant on Civil and Political Reights and the Convention on the Rights of the Child, Iran is obligated to abolish such executions.
Stonings	In 2002, the head of Iran's judiciary issued a ban on stoning. However, Iranian officials later called that directive "advisory" and could be

Table 4. (Continued)

Group/Issue	Regime Practice/Recent Developments
	ignored by individual judges. On December 2, 2008, Iranconfirmed the stoning deaths of two men in Mashhad who were convicted of adultery.
Azeris	Azeris are one quarter of the population, but they complain of ethnic and linguistic discrimination. In 2008, there were several arrests of Azeri students and cultural activists who were pressing for their right to celebrate their culture and history.
Arrests of Dual Nationals	An Iranian-American journalist, Roxanna Saberi, was arrested in January 2009 allegedly because her press credentials had expired; she was charged on April 9, 2009, with espionage, apparently for possessing an Iranian military document. Sentenced to eight years in jail; she was released on appeal on May 12, 2009, but barred from practicing journalism, and has left Iran. Another dual ational, Esha Momeni, arrested in October 2008, is unable to leave Iran. U.S. national, former FBIagent Robert Levinson, remains missing after a visit in 2005 to Kish Island. Iran was given a U.S. letter on these cases at a March 31, 2009, meeting in the Netherlands on Afghanistan.

Sources: Most recent State Department reports on human rights (February 25, 2009), trafficking in persons (June 16, 2009), and on religious freedom (September 19, 2008). www.state.gov

Exiled Opposition Groups

Some groups are committed to the replacement of the regime and remain mostly in exile.

People's Mojahedin Organization of Iran (PMOI)/Camp Ashraf

Of the groups seeking to replace rather than moderate the regime, one of the best known is the People's Mojahedin Organization of Iran (PMOI).[5] Secular and left-leaning, it was formed in the 1960s to try to overthrow the Shah of Iran and advocated Marxism blended with Islamic tenets. It allied with pro-Khomeini forces during the Islamic revolution and supported the November 1979 takeover of the U.S. Embassy in Tehran but was later driven into exile. Even though it is an opponent of Tehran, since the late 1980s the State Department has refused contact with the PMOI and its umbrella organization, the National Council of Resistance (NCR). The State Department designated the PMOI as a foreign terrorist organization (FTO) in October 1997[6] and the NCR was named as an alias of the PMOI in the October 1999 re-designation. The FTO designation was prompted by PMOI attacks in Iran that sometimes kill or injure civilians—although the group does not appear to purposely target civilians. In August 14, 2003, the State Department designated the NCR offices in the United States an alias of the PMOI, and NCR and Justice Department authorities closed down those offices. The regime accuses the group of involvement in the post June 2009 presidential election violence.

The State Department report on international terrorism for 2007 asserts that the organization—and not just a radical element of the organization as the group asserts—was responsible for the alleged killing of seven American defense advisers to the former Shah in 1975-1976. The report again notes the group's promotion of women in its ranks and again emphasizes the group's "cult-like" character, including indoctrination of its members and separation of family members, including children, from its activists. The group's alliance with Saddam Hussein's regime in the 1980s and 1990s has contributed to the U.S. shunning of the organization.

Some advocate that the United States not only remove the group from the FTO list but also enter an alliance with the group against Iran. The FTO designation was up for formal review in October 2008, and, in July 2008, the PMOI formally petitioned to the State Department that its designation be revoked, on the grounds that it renounced any use of terrorism in 2001. However, the State Department announced in mid-January 2009 that the group would remain listed; the next review of that status would be in October 2009.

The group is trying to build on recent legal successes in Europe; on January 27, 2009, the European Union (EU) removed the group from its terrorist group list; the group had been so designated by the EU in 2002. In May 2008, a British appeals court determined that the group should no longer be considered a terrorist organization on the grounds that the British government did not provide "any reliable evidence that supported a conclusion that PMOI retained an intention to resort to terrorist activities in the future." In June 2003, France arrested about 170 PMOI members, including its co-leader Maryam Rajavi (wife of PMOI founder Masoud Rajavi, whose whereabouts are unknown). She was released and remains based in France, and is occasionally received by European parliamentarians and other politicians.

The issue of group members in Iraq is increasingly pressing. U.S. forces attacked PMOI military installations in Iraq during Operation Iraqi Freedom and negotiated a ceasefire with PMOI military elements in Iraq, requiring the approximately 3,400 PMOI fighters to remain confined to their Ashraf camp near the border with Iran. Its weaponry is in storage, guarded by U.S. personnel. In July 2004, the United States granted the Ashraf detainees "protected persons" status under the 4th Geneva Convention, meaning they will not be extradited to Tehran or forcibly expelled as long as U.S. forces have a mandate to help secure Iraq. Another 350 PMOI fighters have taken advantage of an arrangement between Iran and the ICRC for them to return to Iran if they disavow further PMOI activities; none are known to have been persecuted since returning.

The U.S.-led security mandate in Iraq was replaced on January 1, 2009, by a bilateral U.S.-Iraq agreement that limits U.S. flexibility in Iraq. The group fears that, now that Iraqi forces have taken control of the camp, Iraq will expel the group to Iran. The Iraqi government has tried to calm those fears in January 2009 by saying that it would not do so, although it reiterated in late March 2009 that members of the group will be expelled. Some observers say Iraq might move the camp to Iraq's interior, away from the Iran border. The EU "de-listing" might help resolve the issue by causing EU governments to take in those at Ashraf.

Other Armed Groups

Some armed groups are operating in Iran's border areas, and are generally composed of ethnic or religious minorities. One such group is *Jundullah*, composed of Sunni Muslims primarily from the Baluchistan region bordering Pakistan. Since mid-2008, it has conducted several successful attacks on Iranian security personnel, apparently including in May 2009, claiming revenge for the poor treatment of Sunnis in Iran. An armed Kurdish group operating out of Iraq is the Free Life Party, known by its acronym PJAK. PJAK was designated in early February 2009 as a terrorism supporting entity under Executive order 13224, although the designation statement indicated the decision was based mainly on PJAK's association with the Turkish Kurdish opposition group Kongra Gel, also known as the PKK. Another militant group, the "*Ahwazi Arabs*," operates in the largely Arab inhabited areas of southwest Iran, bordering Iraq.

The Son of the Former Shah

Some Iranian exiles, as well as some elites still in Iran, want to replace the regime with a constitutional monarchy led by Reza Pahlavi, the U.S.-based son of the late former Shah and a U.S.-trained combat pilot. However, he does not appear to have large-scale support inside Iran. In January 2001, the Shah's son, who is about 50 years old, ended a long period of inactivity by giving a speech in Washington, DC, calling for unity in the opposition and the institution of a constitutional monarchy and democracy in Iran. He has since broadcast messages into Iran from Iranian exile-run stations in California,[7] and delivered a statement condemning the regime for the post-2009 election crackdown.

Other Activists

Numerous Iranians-Americans in the United States want to see a change of regime in Tehran. Many of them are based in California, where there is a large Iranian-American community, and there are about 25 small-scale radio or television stations that broadcast into Iran. Some organizations, such as The National Iranian American Council (NIAC), are not necessarily seeking influence inside Iran. The mission of NIAC, composed largely of Iranian-Americans, is to promote discussion of U.S. policy and the group has advocated engagement with Iran.

IRAN'S STRATEGIC CAPABILITIES AND WEAPONS OF MASS DESTRUCTION PROGRAMS

Many in the Obama Administration share the Bush Administration's assessment, encapsulated in a March 16, 2006 "National Security Strategy" document, that the United States "may face no greater challenge from a single country than from Iran." This assessment is based largely on Iran's growing weapons of mass destruction (WMD) programs and its ability to exert influence in the region counter to U.S. objectives.[8] Many experts agree that Iran's core national security goals are to protect itself from foreign, primarily U.S., interference or attack, and to exert regional influence that Iran believes is commensurate with its size and concept of nationhood.

Iran's armed forces are extensive but they are widely considered relatively combat ineffective against a well-trained, sophisticated military such as that of the United States or a regional power such as Turkey, but Iran's forces could still cause damage to U.S. forces and allies in the Gulf region. Iranian forces are believed sufficiently effective to deter or fend off conventional threats from Iran's weaker neighbors such as post-war Iraq, Turkmenistan, Azerbaijan, and Afghanistan, but lacking in the logistical ability to project power much beyond Iran's borders.

Conventional Military/Revolutionary Guard/Qods Force

Table 5. The Revolutionary Guard

The IRGC is generally loyal to Iran's hardliners politically and is clearly more politically influential than is Iran's regular military, which is numerically larger but was held over from the Shah's era. As described in a 2009 Rand Corporation study, "Founded by a decree from Ayatollah Khomeini shortly after the victory of the 1978-1979 Islamic Revolution, Iran's Islamic Revolutionary Guards Corps (IRGC) has evolved well beyond its original oundations as an ideological guard for the nascent revolutionary regime. Today the IRGC functions as an expansive socio-political-economic conglomerate whose influence extends into virtually every corner of Iranian political life and society. Bound together by the shared experience of war and the socialization of military service, the Pasdaran have articulated a populist, authoritarian, and assertive vision for the Islamic Republic of Iran that they maintain is a more faithful reflection of the revolution's early ideals. The IRGC's presence is particularly powerful in Iran's highly factionalized political system, in which [many senior figures] hail from the ranks of the IRGC. Outside the political realm, the IRGC oversees a robust apparatus of media resources, training activities, education programs designed to bolster loyalty to the regime, prepare the citizenry for homeland defense, and burnish its own institutional credibility vis-a-vis other factional actors. It is in the economic sphere, however, that the IRGC has seen the greatest growth and diversification – strategic industries and commercial services ranging from dam and pipeline construction to automobile manufacturing and laser eye surgery have fallen under its sway, along with a number of illicit smuggling and black market enterprises."
As further evidence of the IRGC pre-eminence in the conventional command structure, Chairman of the Joint Chiefs of Staff Admiral Mike Mullen said on November 29, 2007, that the IRGC Navy was given responsibility to patrol the entire Persian Gulf, and that the regular Navy is patrolling the Strait of Hormuz and Gulf of Oman.
Through its Qods (Jerusalem) Force, the IRGC has a foreign policy role in exerting influence throughout the region by supporting pro-Iranian movements, as discussed further below. The Qods Force numbers approximately 10,000- 15,000 personnel who provide advice, support, and arrange weapons deliveries to pro-Iranian factions in Lebanon, Iraq, Persian Gulf states, Gaza/West Bank, Afghanistan, and Central Asia. It also operates a worldwide intelligence network to give Iran possible terrorist option and to assist in procurement of WMD-related technology. The Qods Force commander, Brig. Gen. Qassem Soleimani, is said to have his own independent channel to Supreme Leader Khamene'i, bypassing the IRGC and Joint Staff command structure.
IRGC leadership developments are significant because of the political influence of the IRGC. On September 2, 2007, Khamene'i replaced Rahim Safavi with Mohammad Ali Jafari as Commander In Chief of the Guard; Jafari is considered a hardliner against political dissent and is reputedly close to the Supreme Leader and less so to Ahmadinejad. The *Basij* is commanded
by Hossein Taeb but reports to the IRGC Commander in Chief. It operates from thousands of positions in Iran's institutions. Command reshuffles in July 2008 that integrated the Basij more closely with provincially-based IRGC units furthered the view that the Basij is playing a more active role in uncovering suspected plotting by Iran's minorities and others. More information on Iran's military and how it might perform against the United States is discussed under "military options" below.
As noted, the IRGC is also increasingly involved in Iran's economy, acting through a network of contracting businesses it has set up, most notably *Ghorb* (also called *Khatem ol-Anbiya*, Persian for "Seal of the Prophet"). Active duty IRGC senior commanders reportedly serve on Ghorb's board of directors. On October 21, 2007, the Treasury Department designated several IRGC companies as proliferation entities under Executive Order 13382, as shown in the table at the end of the paper. Also that day, the IRGC as a whole, the Ministry of Defense, several IRGC commanders, and several Iranian banks were sanctioned under that same executive order. Simultaneously, the Qods Force was named as a terrorism supporting entity under Executive Order 13224. Both orders freeze the U.S.-based assets and prevent U.S. transactions with the named entities, but these entities are believed to have virtually no U.S.-based assets that could be frozen. The designations stopped short of concurring with provisions of bills in the 110th Congress—H.R. 1400 (passed by the House on September 25), S. 970, and the FY2008 defense authorization bill (P.L. 110-181, Senate amendment adopted September 6, 2007, by vote of 76-22)—for the Revolutionary Guard to be designated a foreign terrorist organization, or FTO.

Source: Frederic Wehrey et al. "The Rise of the Pasdaran." Rand Corporation. 2009. Katzman, Kenneth. "The Warriors of Islam: Iran's Revolutionary Guard." Westview Press, 1993.

Iran's armed forces have few formal relationships with foreign militaries, but Iran and India have a "strategic dialogue" and some Iranian naval officers reportedly have undergone some training in India. Iran and Turkey have agreed in principle (April 2008) to jointly fight terrorism along their border.

Most of Iran's other military-to-military relationships, such as with Russia, Ukraine, Belarus, North Korea, and a few others, generally center on Iranian arms purchases or upgrades.

Iran's armed forces are divided organizationally. The Islamic Revolutionary Guard Corps (IRGC, known in Persian as the *Pasdaran*)[9] controls the *Basij* (mobilization) volunteer militia that enforces adherence to Islamic customs and has been the main instrument to repress the postelection protests in Iran. The IRGC and the regular military report to a Joint Headquarters, headed by Hassan Firuzabadi.

Nuclear Program and Related International Diplomacy

Since 2005, Iran and the international community have been on a "collision course" over Iran's nuclear program. U.S. officials say they are operating under the assumption that Iran intends to develop a nuclear weapon from that program. International Atomic Energy Agency (IAEA) Director Mohammad El Baradei said in a press interview on June 17, 2009, that "My gut feeling is that Iran definitely would like to have the technology...that would enable it to have nuclear weapons if they decided to do so." In February 2009, the Director of National Intelligence (DNI) reiterated previous assessments that it is likely that Iran will eventually try to develop a nuclear weapon.

The Obama Administration faces policy choices as the IAEA reports (February 19, 2009) that Iran has now enriched enough uranium for a nuclear weapon, although only if enriched to 90%. The IAEA report of June 5, 2009, reiterated that Iran's enrichment thus far has been 5%, which is a level that would permit only civilian uses, but added that Iran has now installed over 7,000 centrifuges, of which over 5,000 are being fed with uranium feedstock. There continues to be no evidence that Iran has diverted any nuclear material for a nuclear weapons program, but the IAEA asserts that it cannot verify that Iran's current program is purely peaceful. Several of its reports (January 31, 2006, February 27, 2006, May 26, 2008, and September 15, 2008) describe Iranian documents that show a possible involvement of Iran's military in the program.

Some Administration officials lean toward an alarmist view of Iran's capabilities, while others, including Secretary of Defense Gates, have signaled less urgency, saying on March 1, 2009, that Iran is "not close" to a nuclear weapon. The George W. Bush Administration's December 2007 NIE assessed that Iran will likely be technically capable of producing enough highly enriched uranium for a nuclear weapon some time during 2010-2015. This time frame was reiterated in February 12, 2009, testimony by Director of National Intelligence Dennis Blair and by him again since then, although the DNI reiterated that Iran's weaponization efforts appear to still be on hold since 2003, as the 2007 NIE had said. [10]

Table 6. Iran's Conventional Military Arsenal

Military Personnel	Tanks	Surface-Air Missiles	Combat Aircraft	Ships	Defense Budget (billions U.S. $)
545,000 (regular military and Revolutionary Guard Corps (IRGC). IRGC is about one-third of total force.	1,693 (incl. 480 T-72)	150 I-Hawk plus some Stinger	280 (incl. 25 MiG-29 and 30 Su-24)	200 (incl. 10 Chinese-made Hudong, 40 Boghammer, 3 frigates) Also has 3 Kilo subs	6.6
Ship-launched cruise missiles. Iran is able to arm its patrol boats with Chinese-made C-802 cruise missiles. Iran also has Chinese-supplied HY-2 Seerseekers emplaced along Iran's coast.					
Midget Subs. Iran is said to possess several, possibly purchased assembled or in kit form from North Korea. Iran claimed on Nov. 29, 2007 to have produced a new small sub equipped with sonar-evading technology.					
Anti-aircraft missile systems. Russia has sold and now delivered to Iran (January 2007) 30 anti-aircraft missile systems (Tor M1), worth over $1 billion. In September 2006, Ukraine agreed to sell Iran the Kolchuga radar system that can improve Iran's detection of combat aircraft. In December 2007, Russia agreed to sell the even more capable S-300 (also known as SA-20 "Gargoyle") air defense system, purportedly modeled after the U.S. Patriot system, which U.S. officials say would greatly enhance Iran's air defense capability. The value of the deal is estimated at $800 million. Amid unclear or weak denials by Iranian and Russian officials, U.S. officials told journalists on December 11, 2008, that Iran has indeed contracted for the missile. It is reportedly was due for delivery by March 2009 and to be operationalby June 2009, but Russian press reports in February 2009 about the visit of Iran's Defense Minister to Moscow indicate that Russia has placed delivery on hold due to "political considerations"—expectations of possible adverse reaction by the Obama Administration. Delivery has not taken place to date, by all accounts.					

Iran's Arguments and the International Response

International scrutiny of Iran's nuclear program intensified in 2002, when Iran confirmed PMOI allegations that Iran was building two facilities that could potentially be used to produce fissile material useful for a nuclear weapon: a uranium enrichment facility at Natanz and a heavy water production plant at Arak,[11] considered ideal for the production of plutonium. It was revealed in 2003 that the founder of Pakistan's nuclear weapons program, A.Q. Khan, sold Iran nuclear technology and designs.[12]

Iranian leaders have answered the criticism saying that Iran's nuclear program is for electricity generation and that enrichment is its "right" as a party to the NPT. Iran professes that WMD is inconsistent with its ideology and says that its leaders, including the late Ayatollah Khomeini, have issued formal pronouncements (*fatwas*) that nuclear weapons are un-Islamic. Iran says its oil resources are finite and that enriching uranium to make nuclear fuel is allowed under the 1968 Nuclear Non-Proliferation Treaty,[13] to which Iran is a party. An analysis was published by the National Academy of Sciences challenging the U.S. view that Iran is petroleum rich and therefore has no need for a nuclear power program. According to the analysis, the relative lack of investment could cause Iran to have negligible exports of oil by 2015.[14]

The United States and its partners accept Iran's right to purely peaceful uses of nuclear energy, if that is truly Iran's intent. U.S. officials say that Iran's gas resources make nuclear energy unnecessary. However, U.S. officials assess that Iran's professions of a purely peaceful nuclear program cannot be trusted because Iran's factions appear to agree on the utility of a nuclear weapons capability as a means of ending its perceived historic vulnerability to invasion and domination by great powers, and as a symbol of Iran as a major nation. Others believe a nuclear weapon represents the instrument with which Iran intends to intimidate its neighbors and dominate the Persian Gulf region. There are also fears Iran might transfer WMD to extremist groups or countries.

U.S. officials have generally been less concerned with Russia's work, under a January 1995 contract, on an $800 million nuclear power plant at Bushehr. Russia insisted that Iran sign an agreement under which Russia would provide reprocess the plant's spent nuclear material; that agreement was signed on February 28, 2005. The plant was expected to become operational in 2007, but Russia had insisted (including during President Putin's visit to Iran in October 2007) that Iran first comply with the U.N. resolutions discussed below. In December 2007, perhaps to signal disagreement with further pressure on Iran, Russia began fueling the reactor, and Iran says it expects the plant to become operational in 2009. Some preliminary tests of the plant began in February 2009. As part of its work with Iran, Russia has trained 1,500 Iranian nuclear engineers.

Diplomatic Efforts in 2003 and 2004/Paris Agreement

In 2003, France, Britain, and Germany (the "EU-3") opened a separate diplomatic track to curb Iran's program. On October 21, 2003, Iran pledged, in return for peaceful nuclear technology, to (1) fully disclose its past nuclear activities, (2) to sign and ratify the "Additional Protocol" to the NPT (allowing for enhanced inspections), and (3) to suspend uranium enrichment activities. Iran signed the Additional Protocol on December 18, 2003, although the *Majles* has not ratified it. Iran discontinued abiding by the Protocol after the IAEA reports of November 10, 2003, and February 24, 2004, stated that Iran had violated its NPT reporting obligations over an 18-year period.

In the face of the U.S. threat to push for Security Council action, the EU-3 and Iran reached a more specific November 14, 2004, "Paris Agreement," committing Iran to suspend uranium enrichment (which it did as of November 22, 2004) in exchange for renewed trade talks and other aid.[15] The Bush Administration did not openly support the track until March 11, 2005 when it announced it would drop U.S. objections to Iran applying to join the World Trade Organization (it applied in May 2005) and to selling civilian aircraft parts to Iran. The Bush Administration said it would not participate directly in the talks.

Reference to the Security Council

The Paris Agreement broke down just after Ahmadinejad's election; Iran rejected as insufficient an EU-3 offer to assist Iran with peaceful uses of nuclear energy and provide limited security guarantees in exchange for Iran's (1) permanently ending uranium enrichment; (2) dismantling the Arak heavy water reactor;[16] (3) no-notice nuclear inspections; and (4) a pledge not to leave the NPT (it has a legal exit clause).

Table 7. Summary of Provisions of U.N. Resolutions on Iran Nuclear Program (1737, 1747, and 1803)

Require Iran to suspend uranium enrichment
Prohibit transfer to Iran of nuclear, missile, and dual use items to Iran, except for use in light water reactors
Prohibit Iran from exporting arms or WMD-useful technology
Freeze the assets of 40 named Iranian persons and entities, including Bank Sepah, and several Iranian front companies
Require that countries exercise restraint with respect to travel of 35 named Iranians and ban the travel of 5 others
Calls on states not to export arms to Iran or support new business with Iran
Calls for vigilance with respect to the foreign activities of all Iranian banks, particularly Bank Melli and Bank Saderat
Calls on countries to inspect cargoes carried by Iran Air Cargo and Islamic Republic of Iran Shipping Lines if there are indications they carry cargo banned for carriage to Iran.

On August 8, 2005, Iran broke the IAEA seals and began uranium "conversion" (one step before enrichment) at its Esfahan facility. On September 24, 2005, the IAEA Board declared Iran in non-compliance with the NPT and decided to refer the issue to the Security Council,[17] but no time frame was set for the referral. After Iran resumed enrichment activities, on February 4, 2006, the IAEA board voted 27-3[18] to refer the case to the Security Council. On March 29, 2006, the Council agreed on a presidency "statement" setting a 30-day time limit (April 28, 2006) for ceasing enrichment.[19]

Establishment of "P5+1" Contact Group/June 2006 Incentive Package

Taking a multilateral approach, the George W. Bush Administration offered on May 31, 2006, to join the nuclear talks with Iran if Iran first suspends its uranium enrichment. Such talks would center on a package of incentives and possible sanctions—formally agreed on June 1, 2006—by a newly formed group of nations, the so-called "Permanent Five Plus 1" (P5+1: United States, Russia, China, France, Britain, and Germany). EU representative Javier Solana formally presented the P5+1 offer to Iran on June 6, 2006. (The package is Annex I to Resolution 1747.)

Incentives:

- Negotiations on an EU-Iran trade agreements and acceptance of Iran into the World Trade Organization.
- Easing of U.S. sanctions to permit sales to Iran of commercial aircraft/parts.
- Sale to Iran of a light-water nuclear reactor and guarantees of nuclear fuel (including a five year buffer stock of fuel), and possible sales of light-water research reactors for medicine and agriculture applications.
- An "energy partnership" between Iran and the EU, including help for Iran to modernize its oil and gas sector and to build export pipelines.
- Support for a regional security forum for the Persian Gulf, and support for the objective of a WMD free zone for the Middle East.

- The possibility of eventually allowing Iran to resume uranium enrichment if it complies with all outstanding IAEA requirements.

Reported Sanctions:[20]

- Denial of visas for Iranians involved in Iran's nuclear program and for high-ranking Iranian officials.
- A freeze of assets of Iranian officials and institutions; a freeze of Iran's assets abroad; and a ban on some financial transactions.
- A ban on sales of advanced technology and of arms to Iran; and a ban on sales to Iran of gasoline and other refined oil products.
- An end to support for Iran's application to the WTO.

Resolution 1696

Iran did not immediately respond to the incentive offer. On July 31, 2006, the Security Council voted 14-1 (Qatar voting no) for U.N. Security Council Resolution 1696, giving Iran until August 31, 2006, to fulfill the longstanding IAEA nuclear demands (enrichment suspension, etc). Purportedly in deference to Russia and China, it was passed under Article 40 of the U.N. Charter, which makes compliance mandatory, but not under Article 41, which refers to economic sanctions, or Article 42, which would authorize military action. It called on U.N. member states not to sell Iran WMD-useful technology. On August 22, 2006, Iran formally responded to the June proposal, but Iran did not offer enrichment suspension, instead proposing broader engagement with the West and guarantees that the United States would not seek regime change.

Resolution 1737

With the backing of the P5+1, chief EU negotiator Javier Solana negotiated with Iran to try arrange a temporary enrichment suspension, but talks ended on September 28, 2006, without agreement. The Security Council adopted U.N. Security Council Resolution 1737 unanimously on December 23, 2006, under Chapter 7, Article 41 of the U.N. Charter. It prohibits sale to Iran—or financing of such sale—of technology that could contribute to Iran's uranium enrichment or heavy-water reprocessing activities. It also required U.N. member states to freeze the financial assets of 10 named Iranian nuclear and missile firms and 12 persons related to those programs (see Table 10). It called on – but did not mandate -- member states not to permit travel by these persons. In deference to Russia, the Resolution did not apply to the Bushehr reactor.

Resolution 1747 and Results

Resolution 1737 demanded enrichment suspension by February 21, 2007. With no sign of Iran doing so, on March 24, 2007, after only three weeks of P5+1 negotiations, Resolution 1747 was adopted unanimously, which:

- added 10 military/WMD-related entities; 3 Revolutionary Guard entities; 8 persons, and 7 Revolutionary Guard commanders listed in Table 10. Bank Sepah is among the entities sanctioned.

- banned arms transfers by Iran, a provision targeted at Iran's alleged arms supplies to Lebanese Hezbollah and to Shiite militias in Iraq.
- required all countries to report to the United Nations when sanctioned Iranian persons travel to their territories.
- called for (but did not require) countries to avoid selling arms or dual use items to Iran and to avoid any new lending or grants to Iran.

Resolution 1747 demanded Iran suspend enrichment by May 24, 2007. Iran did not comply, but, suggesting it wanted to avoid further isolation, in August 2007, Iran agreed to sign with the IAEA an agreement to clear up outstanding questions on Iran's past nuclear activities by the end of 2007. On September 28, 2007, the P5+1 grouping—along with the EU itself—agreed to a joint statement pledging to negotiate another sanctions resolution if there is no progress reported by the IAEA in implementing the August 2007 agreement or in negotiations with EU representative Javier Solana. The IAEA and Solana indicated that Iran's responses fell short; Solana described a November 30, 2007, meeting with Iranian negotiator Sayid Jallili as "disappointing."

Resolution 1803 and Additional Incentives

After several months of negotiations, Resolution 1803 was adopted by a vote of 14-0 (Indonesia abstaining) on March 3, 2008. It: (1) bans sales of dual use items to Iran; (2) authorizes, but does not require, inspections of cargo (carried by Iran Air Cargo and Islamic Republic of Iran Shipping Line) suspected of shipping WMD-related goods; (3) imposes a firm travel ban on five Iranians named in Annex II to the Resolution and requires reports on international travel by 13 individuals named in Annex I; (4) calls for, but does not impose, a prohibition on financial transactions with Iran's Bank Melli and Bank Saderat; and (5) adds 12 entities to those sanctioned under Resolution 1737. Although these provisions did not sanction civilian trade, the United States hailed the Resolution as demonstrating international unity. (On June 23, 2008, the EU, acting under Resolution 1803, froze the assets of Bank Melli and several IRGC entities and commanders.)

Resolution 1803 also stated that "China, France, Germany, the Russian Federation, the United Kingdom and the United States are willing to take further concrete measures on exploring an overall strategy of resolving the Iranian nuclear issue through negotiation on the basis of their June 2006 proposals." Trying to preserve a unified front, the Bush Administration agreed to expand the June 2006 incentive package to induce Iranian cooperation. At a meeting in London on May 2, 2008, the P5+1 agreed on a "refreshed" package of incentives to augment those in the June 2006 package. According to press reports (the exact offer was not made public), the powers offered, beyond previous offers, political cooperation with Iran, and enhanced incentives on energy cooperation. EU envoy Solana presented the package (which included a signature by Secretary of State Rice) on June 14, 2008, but Iran was non-committal.

Sensing increasing pressure, Iranian Foreign Minister Mottaki indicated on July 2, 2008, that Iran might be ready to negotiate on the June 2008 incentive package by first accepting a six week "freeze for freeze"— the P5+1 would freeze further sanctions efforts and Iran would freeze any expansion of uranium enrichment (though not suspend outright). To try to take advantage of what seemed to be divisions within Iran on whether to negotiate a settlement, the Administration sent Undersecretary of State for Political Affairs William Burns to join Solana and the other P5+1 representatives at a meeting in Geneva on July 19, 2008, to receive Iran's response to the "freeze for freeze" idea. Iran did not supply a direct answer either then, or by an extended deadline of August 2, 2008, instead focusing on what would be contained in a final nuclear settlement.

Resolution 1835

As a result of the lack of progress, the P5+1 began discussing another sanctions resolution. Ideas reportedly under consideration included adding more Iranian banks to those with which business would be banned, or banning insurance for Iran's tanker fleet. On August 7, 2008, separately from any U.N. action but intended to increase pressure on Iran, the EU not only implemented those sanctions specified in Resolution 1803, including the authority to inspect suspect shipments, but also called on its member states to refrain from providing new credit guarantees on exports to Iran. However, the August 2008 crisis between Russia and Georgia set back U.S.-Russia relations, and Russia opposed new sanctions on Iran. In an effort to demonstrate to Iran continued unity, the Council did adopt (September 27, 2008) Resolution 1835, calling on Iran to comply with previous resolutions, but restating a willingness to negotiate and imposing no new sanctions.

With Iran still not complying, the P5+1 met again on October 21, 2008, and in Paris on November 13, 2008. However, with U.S. partner officials uncertain about what U.S. policy toward Iran might be under a new U.S. Administration, there was no consensus on new sanctions.

The P5+1 Process under President Obama

After President Obama was inaugurated, the P5+1 met in Germany (February 4, 2009), reportedly focusing on how the new U.S. Administration planned to change the U.S. approach on Iran. The P5+1 countries issued a statement recommitting to the "two track" strategy of incentives and increasing sanctions.[21] Another P5+1 meeting was held in London on April 8, 2009, during which Undersecretary Burns told the other members of the group that, henceforth, a U.S. diplomat would attend all of the group's meetings with Iran. Iran said any new meetings with the P5+1 would have to await the Iranian June 12, 2009 election. The Obama Administration did not, at these meetings, press for new sanctions, pending indications of how Iran would respond to its diplomatic outreach efforts. Some reports in April 2009 said the Administration was considering dropping the precondition that Iran suspend uranium enrichment before talks on a final status of Iran's program would begin. Iranian leaders said in April 2009 that they are considering offering new proposals to resolve the nuclear issues, but no specifics have been announced to date. On May 18, 2009, in the context of a meeting with visiting Israeli Prime Minister Benjamin Netanyahu, President Obama said he expect a positive response by Iran to the U.S. outreach "by the end of [2009]," but that the United States would not entertain the idea of endless talks that yield no result on Iran's nuclear program. Most observers say the Iranian election dispute is likely to significantly delay any new negotiations with Iran on the issue and could prompt an early return to P5+1 discussions of new sanctions on Iran, although some believe Iran will try to restore its international legitimacy through new flexibility on this issue.

Some outside experts have floated compromise proposals. Former senior U.S. diplomat Thomas Pickering and other experts said in April 2008 that U.S. and Iranian former officials and academics have been meeting to discuss formulas under which Iran might continue to enriched uranium to non bomb-grade levels under monitoring.

Chemical Weapons, Biological Weapons, and Missiles

Official U.S. reports and testimony continue to state that Iran is seeking a self-sufficient chemical weapons (CW) infrastructure, and that it "may have already" stockpiled blister, blood, choking, and nerve agents—and the bombs and shells to deliver them. This raises questions about Iran's compliance with its obligations under the Chemical Weapons Convention (CWC), which Iran signed on January 13, 1993, and ratified on June 8, 1997. These officials and reports also say that Iran "probably maintain[s] an offensive [biological weapons] BW program ... and probably has the capability to produce at least small quantities of BW agents."

Ballistic Missiles/Warheads

At the 2009 Annual Threat Assessment of the Intelligence Community, Director of National Intelligence, Dennis Blair, testified "although many of their statements are exaggerations, Iranian officials throughout the past year have repeatedly claimed greater ballistic missile capabilities that could threaten U.S. and allied interests."[22] Tehran appears to view its ballistic missiles as an integral part of its strategy to deter or retaliate against forces in the region, including U.S. forces. To reinforce that message, during July 9-10, 2008, Iran conducted several highly publicized test launches of a variant of the Shahab missile that Iran says has a 1,250 mile range, putting most key U.S. allies in the region at risk; other missiles were tested on those days as well. Some analyses suggested Iran might have falsified photos of some of the missiles.

However, Iran's technical capabilities are a matter of some debate among experts. As of August 19, 2008, the George W. Bush Administration has reached agreements with Poland and the Czech Republic to establish a missile defense system to counter Iranian ballistic missiles. These agreements were reached over Russia's opposition, which was based on the belief that the missile defense system would be used to neutralize Russian capabilities. The Obama Administration has de-emphasize this program in the context of efforts to improve relations with Russia.

Foreign Policy and Support for Terrorist Groups

Iran's foreign policy is a product of the ideology of Iran's Islamic revolution, blended with longstanding national interests. Some interpret Iran's objectives as the overturning of the power structure in the Middle East that Iran believes favors the United States, Israel, and their "collaborators"—Sunni Muslim regimes such as Egypt, Jordan, and Saudi Arabia.

Table 8. Iran's Ballistic Missile Arsenal

Shahab-3 ("Meteor")	800 mile range. Two of first three tests (July 1998, July 2000, and September 2000) reportedly unsuccessful. After successful test in June 2003, Iran called missile operational (capable of hitting Israel). Despite claims, some U.S. experts say the missile not completely reliable—some observers said Iran detonated in mid-flight a purportedly more accurate version on August 12, 2004. On May 31, 2005, Iran announced it had tested a solid-fuel version.
Shahab-4 / Sijjil	1,200 - 1,500 mile range. In October 2004, Iran announced it had extended range of the Shahab-3 to 1,200 miles, and it added in early November 2004 that it is capable of "mass production" of it. *Agence France Presse* report (February 6, 2006) said January 2006 test succeeded. Related missiles claimed by Iran to have 1,200 mile range, include the "Ashoura" (claimed in November 2007); the "Ghadr" (displayed at military parade in September 2007); and the "Sijil," tested on November 12, 200, (solid fuel). "Sijil 2" tested successfully on May 20, 2009, but Secretary Gates said the range is likely closer to 1,200 miles than to 1,500. Still, this test potentially puts large portions of the Near East and Southeastern Europe in range, including U.S. bases in Turkey.
BM-25	1,500 mile range. On April 27, 2006, Israel's military intelligence chief said that Iran had received a shipment of North Korean-supplied BM-25 missiles. Missile said to be capable of carrying nuclear warheads. The *Washington Times* appeared to corroborate this reporting in a July 6, 2006, story, which asserted that the North Korean-supplied missile is based on a Soviet-era "SS-N-6" issile.
ICBM	U.S. officials believe Iran might be capable of developing an intercontinental ballistic missile (3,000 mile range) by 2015. In February 2008 Iran claimed to have launched a probe into space, suggesting its missile technology might be improving to the point where an Iranian ICBM is realistic.
Other Missiles	On September 6, 2002, Iran said it successfully tested a 200 mile range "Fateh 110" missile (solid propellent), and Iran said in late September 2002 that it had begun production. Iran also possesses a few hundred short-range ballistic missiles, including the *Shahab-1* (Scud-b), the *Shahab-2* (Scud-C), and the *Tondar-69* (CSS-8). In January 2009, Iran claimed to have tested a new air-to-air missile.
Shahab-3 ("Meteor")	800 mile range. Two of first three tests (July 1998, July 2000, and September 2000) reportedly unsuccessful. After successful test in June 2003, Iran called missile operational (capable of hitting Israel). Despite claims, some U.S. experts say the missile not completely reliable—some observers said Iran detonated in mid-flight a purportedly more accurate version on August 12, 2004. On May 31, 2005, Iran announced it had tested a solid-fuel version.
Space Vehicle	Following an August 2008 failure, in early February 2009, Iran successfully launched a small, low- earth satellite on a Safir-2 rocket (range about 155 miles). The Pentagon said the launch was "clearly a concern of ours" because "there are dual-use capabilities here which could be applied toward the development of long-range missiles."
Warheads	*Wall Street Journal* report of September 14, 2005, said that U.S. intelligence believes Iran is working to adapt the Shahab-3 to deliver a nuclear warhead. Subsequent press reports say that U.S. intelligence captured an Iranian computer in mid-2004 showing plans to construct a nuclear warhead for the Shahab.[23] The IAEA is seeking additional information from Iran.

The State Department report on international terrorism for 2008 released April 30, 2009, again stated (as it has for more than a decade) that Iran "remained the most active state sponsor of terrorism" in 2008, and it again attributed the terrorist activity primarily to the Qods Force of the Revolutionary Guard. The report focused particular attention on Iran's lethal support to Shiite militias in Iraq as well as on shipments to and training of "selected" Taliban fighters in Afghanistan.[24] On October 27, 2008, the deputy commander of the Basij became the first top Guard leader to publicly acknowledge that Iran supplies weapons to "liberation armies" in the region, a reference to pro-Iranian movements discussed below.

Some experts believe that Iran is ascendant in the region because of the installation of pro-Iranian regimes in Iraq and Afghanistan, and the strength of Hezbollah in Lebanon and Hamas in Gaza. Iran might, according to this view, seek to press its advantage to strengthen regional Shiite movements and possibly drive the United States out of the Gulf. During a visit to the Middle East in March 2009, Secretary of State Clinton said, after meeting with several Arab and Israeli leaders in the region, that "There is a great deal of concern about Iran from this whole region ... "

Others reach an opposite conclusion, stating that Iran now feels more encircled than ever by pro-U.S. regimes. Recent elections in Lebanon, U.S. engagement with Syria, stability in Iraq, and an influx of U.S. troops to Afghanistan have rendered Iran weaker than it has been in recent years. Some reports say that countries in the region privately welcome Iran's post-election turmoil because it means that Iran might be strategically weakened and consumed by internal infighting. Others say Iran is more likely to assert itself in the region to focus public attention on foreign and regional threats.

Relations with the Persian Gulf States

The Persian Gulf monarchy states (Gulf Cooperation Council, GCC: Saudi Arabia, Kuwait, Bahrain, Qatar, Oman, and the United Arab Emirates) fear the growing strategic influence of Iran but they do not openly support U.S. conflict with Iran that might cause Iran to retaliate against Gulf state targets. At the same time, since the mid-1990s, Iran has tried to blunt Gulf state fears of Iran by curtailing activity, conducted during the 1980s and early 1990s, to sponsor Shiite Muslim extremist groups in these states, all of which are run by Sunni governments. Iran found, to its detriment, that subversion caused the Gulf states to ally closely with the United States. In part to counter Iran's perceived growing influence in the Gulf, in December 2006 the summit of the GCC leaders announced that the GCC states might jointly study their own development of "peaceful nuclear technology." On the other hand, seeking to avoid further tensions with Iran, the GCC leaders invited Ahmadinejad to speak at the December 2-3, 2007, summit of the GCC leaders in Doha, Qatar, marking the first time an Iranian president has been invited since the GCC was formed in 1981. His speech reiterated a consistent Iranian theme that the Gulf countries, including Iran, should set up their own security structure without the help of "outside powers" but also called for a "new chapter" in Iran-GCC relations.

- *Saudi Arabia.* Many observers closely watch the relationship between Iran and Saudi Arabia because of Saudi alarm over the emergence of a pro-Iranian government in

Iraq and Iran's ascendancy in Lebanon. Saudi Arabia sees itself as leader of the Sunni Muslim world and views Shiite Muslims as heretical and disloyal internally. The Saudis, who do not want a repeat of Iran's sponsorship of disruptive and sometimes violent demonstrations at annual Hajj pilgrimages in Mecca in the 1980s and 1990s—or an increase in Iranian support for Saudi Shiite dissidents—are receptive to easing tensions with Iran. They hosted Ahmadinejad in the Kingdom in March 2007 and again for the Hajj in December 2007. The Saudis continue to blame a pro-Iranian movement in the Kingdom, Saudi Hezbollah, for the June 25, 1996, Khobar Towers housing complex bombing, which killed 19 U.S. airmen.[25] After restoring relations in December 1991 (after a four-year break), Saudi-Iran ties progressed to high-level contacts during Khatemi's presidency, including Khatemi visits in 1999 and 2002.

- *United Arab Emirates* (UAE) concerns about Iran never fully recovered from the April 1992 Iranian expulsion of UAE security forces from the Persian Gulf island of Abu Musa, which it and the UAE shared under a 1971 bilateral agreement. (In 1971, Iran, then ruled by the U.S.-backed Shah, seized two other islands, Greater and Lesser Tunb, from the emirate of Ras al-Khaymah, as well as part of Abu Musa from the emirate of Sharjah.) The UAE (particularly the federation capital, Abu Dhabi, which takes a harder line than Dubai, which has an Iranian-origin resident community as large as 300,000 and business ties to Iran) wants to refer the dispute to the International Court of Justice (ICJ), but Iran insists on resolving the issue bilaterally. The UAE has not pressed the issue vigorously in recent years, although it formally protested Iran's setting up of a maritime and ship registration office on Abu Musa in July 2008. The UAE insists the islands dispute be kept on the agenda of the U.N. Security Council (which it has been since December 1971). The United States supports UAE proposals but takes no formal position on sovereignty. As an indicator of the degree to which the issue has faded, in May 2007 the UAE received Ahmadinejad (the highest level Iranian visit since the 1979 revolution) and allowed him to lead an anti-U.S. rally of a reported several hundred Iranian-origin residents of Dubai at a soccer stadium there. On the other hand, tensions have flared since late 2008 over UAE's policy of fingerprinting of Iranian visitors.
- *Qatar*, like most of the other Gulf states, does not seek confrontation and seeks to accommodate some of its interests, yet Qatar remains wary that Iran might eventually seek to encroach on its large North Field (natural gas). It shares that field with Iran (called South Pars on Iran's side) and Qatar earns large revenues from natural gas exports from it. Qatar's fears were heightened on April 26, 2004, when Iran's deputy Oil Minister said that Qatar is probably producing more gas than "her right share" from the field and that Iran "will not allow" its wealth to be used by others. These concerns might have prompted Qatar to invite Ahmadinejad to the December 2007 GCC summit in Qatar.
- *Bahrain* is about 60% Shiite, many of whom are of Persian origin, but its government is dominated by the Sunni Muslim Al Khalifa family. In 1981 and again in 1996, Bahrain publicly accused Iran of supporting Bahraini Shiite dissidents (the Islamic Front for the Liberation of Bahrain, Bahrain-Hezbollah, and other Bahraini dissident groups) in efforts to overthrow the ruling Al Khalifa family. Bahraini fears that Iran would try to interfere in Bahrain's November 25, 2006, parliamentary elections by

providing support to Shiite candidates did not materialize, although the main Shiite opposition coalition won 18 out of the 40 seats of the elected body. Tensions have flared several times since July 2007 when Iranian editorialists asserted that Bahrain is part of Iran—that question was the subject of the 1970 U.N.-run referendum in which Bahrainis opted for independence. The issued flared again after a February 20, 2009, statement by Ali Akbar Nateq Nuri, an adviser to Khamene'i, that Bahrain was at one time an Iranian province. The statement caused major criticism of Iran throughout the region, and contributed to a decision by Morocco to break relations with Iran. Still, Bahrain has sought not to antagonize Iran and has apparently allowed Iran's banks to establish a presence in Bahrain's vibrant banking sector. On March 12, 2008, the Treasury Department sanctioned the Bahrain-based Future Bank under Executive order 13382 that sanctions proliferation entities. Future Bank purportedly is controlled by Bank Melli.

- *Oman*. Of the GCC states, the Sultanate of Oman is closest politically to Iran and has refused to ostracize or even harshly criticize Iranian policies. Some press reports say local Omani officials routinely turn a blind eye to or even cooperate in the smuggling of western goods to Iran.

Iranian Policy in Iraq

The U.S. military ousting of Saddam Hussein benefitted Iran strategically,[26] and during 2004-2008, U.S.-Iran differences in Iraq widened to the point where some were describing the competition as a U.S.-Iran "proxy war" inside Iraq. The acute source of tension was evidence, detailed by U.S. commanders in Iraq, that the *Qods* Force was providing arms (including highly lethal "explosively forced projectiles," EFPs, that have killed U.S. soldiers in Iraq), training, guidance, and financing to "special groups" of Shiite militias involved in sectarian violence. The *New York Times* reported on May 5, 2008, that Lebanese Hezbollah militants, who are Arabs, were providing some of the training to the Iraq militants at training camps near Tehran.

However, recent Defense Department reports on Iraq stability corroborate a widespread perception that Iranian interference in Iraq has lessened, including fewer Iranian weapons shipments. This interference is likely to lessen further if Iranian leaders become embroiled in a long term power struggle stemming from the disputed election in Iran. In Iraq itself, the Shiite militias and political parties that benefit most from Iranian support fared poorly in the January 31, 2009, provincial elections in Iraq, and the results were viewed as a setback for Iran's influence in Iraq. Iran also was unable to derail the U.S.-Iraq defense pact (which took effect January 1, 2009). In January 2009, Iraqi Prime Minister Nuri al-Maliki visited Iran for the fourth time since he became Prime Minister, reportedly to assure Iran that the pact did not threaten Iran.

Iran also has signed a number of agreements with Iraq on transportation, energy cooperation, free flow of Shiite pilgrims, border security, intelligence sharing, and other cooperation; several more agreements, including a $1 billion credit line for Iranian exports to Iraq, were signed during Ahmadinejad's March 2-3, 2008, visit to Iraq; implementing agreements were signed in April 2008. The two countries now do about $4 billion in bilateral trade. Additional Iranian credits and trade issues were reportedly discussed during a March 2009 visit to Iran by Iraqi President Jalal Talabani.

After at first rejecting dialogue with Iran on the Iraq issue, the George W. Bush Administration supported and attended several regional conferences on Iraq, attended by Iran, and undertook bilateral talks with Iran on the Iraq issue. The first such meeting, in Baghdad, was on May 28, 2007; the two sides met at the home of Prime Minister Maliki, who opened the meeting. According to then U.S. Ambassador to Iraq Ryan Crocker (the Iranian side was represented by the Iranian Ambassador to Iraq), the two sides agreed on broad principles for Iraq's political evolution and stability, but the United States would judge the dialogue by Iran's cessation of supplies to Shiite militias. Another round of talks was held on July 24, 2007, resulting in an agreement to establish a working group to discuss ways to stabilize Iraq. This group met for the first time on August 6, 2007. Because of signs that Iran had slowed weapons flows into Iraq, another round of talks was tentatively scheduled for December 18, 2007, but Iran repeatedly postponed the talks because of differences over the agenda and the level of talks (Iran wanted them to be at the ambassador level). On May 5, 2008, Iran indefinitely suspended this dialogue, and, in February 2009, Iran said there was no need to resume it.

A provision of the FY2008 defense authorization bill (P.L. 110-181) required a report to Congress on Iran's interference in Iraq. On January 9, 2008, the Treasury Department took action against suspected Iranian and pro-Iranian operatives in Iraq by designating four individuals and one organization as a threat to stability in Iraq under the July 17, 2007 Executive Order 13438, which freezes the assets and bans transactions with named individuals. The named entities, which includes a senior Qods Forces leader, are in the tables on sanctioned entities.

Supporting Palestinian Militant Groups

Iran's support for Palestinian militant groups has long concerned U.S. Administrations, as part of an apparent effort by Tehran to obstruct an Israeli-Palestinian peace that Iran believes would strengthen the United States and Israel. Ahmadinejad's various statements on Israel were discussed above, although Supreme Leader Khamene'i has repeatedly called Israel a "cancerous tumor." He used that term again during a March 4, 2009, press conference in Tehran. In December 2001, purported "moderate" Rafsanjani said that it would take only one Iranian nuclear bomb to destroy Israel, whereas a similar strike against Iran by Israel would have far less impact because Iran's population is large. Iran has hosted numerous conferences to which anti-peace process terrorist organizations were invited (for example: April 24, 2001, and June 2-3, 2002). During his presidency, Khatemi also issued sharp criticisms and recriminations against Israel, but he also conversed with Israel's president at the 2005 funeral of Pope John Paul II. The formal position of the Iranian Foreign Ministry, considered a bastion of moderates, is that Iran would not seek to block an Israeli-Palestinian settlement but that the peace process is too weighted toward Israel to result in a fair settlement.

Iran and Hamas

The State Department report on terrorism for 2007 (mentioned above) again accused Iran of providing "extensive" funding, weapons, and training to Hamas, Palestinian Islamic Jihad (PIJ), the Al Aqsa Martyr's Brigades, and the Popular Front for the Liberation of Palestine-General Command (PFLP-GC). All are named as foreign terrorist organizations (FTO) by the State Department for their use of violence to undermine the Arab-Israeli peace process. Some saw Iran's regional policy further strengthened by Hamas's victory in the January 25, 2006, Palestinian legislative elections, and even more so by Hamas's June 2007 armed takeover of the Gaza Strip. The Hamas gains potentially positioned it to block any peace settlement with Israel. Hamas activists downplay Iranian influence on them, asserting that Iran is mostly Shiite, while Hamas members are Sunni Muslims.[27] Hamas was reputed to receive about 10% of its budget in the early 1990s from Iran, although since then Hamas has cultivated funding from wealthy Persian Gulf donors and supporters in Europe and elsewhere.

Still, it was evident from the December 27, 2008, - January 17, 2009, Israel-Hamas war in Gaza, that Iran provides material support to Hamas. Joint Chiefs Chairman Adm. Mike Mullen said on January 27, 2009, that the United States boarded but did not seize a ship carrying light arms to Hamas from Iran; the ship (the Monchegorsk) later went to Cyprus. On March 11, 2009, a U.N. committee monitoring Iran's compliance with Resolution 1747, which bans Iranian arms exports, said Iran might have violated that resolution with the alleged shipment. Hamas appeared to corroborate allegations of Iranian weapons supplies when its exiled leader, Khaled Meshal, on February 1, 2009, publicly praised Iran for helping Hamas achieve "victory" over Israel in the conflict.[28] On December 29, 2008, Khamene'i said that Muslims worldwide were "duty-bound" to defend Palestinians in the Gaza Strip against the Israeli offensive against the Hamas-run leadership there, but the Iranian leadership did not attempt to send Iranian volunteers to Gaza to fight on Hamas' behalf. Iranian weaponry might also have been the target of a January 2009 strike on a weapons delivery purportedly bound for Gaza in transit via Sudan (presumably via Egypt).

Sunni Arab leaders in Egypt, Jordan, Saudi Arabia, and throughout the region apparently fear Iran's growing regional influence and reported Iranian support for efforts to discredit these leaders for what Iran considers insufficient support for Hamas in its recent war with Israel. Some Iranian efforts reportedly involve establishing Hezbollah cells in some of these countries, particularly Egypt, purportedly to stir up opposition to these governments and build public support for Hezbollah and Hamas. [29] These countries are also said to fear that President Obama's outreach to Iran might lead the United States to downplay their concerns about Iran—a sentiment that Secretary of Defense Gates tried to allay during his visit to the Middle East in May 2009.

Lebanese Hezbollah and Syria

Iran has maintained a close relationship with Hezbollah since the group was formed in 1982 by Lebanese Shiite clerics who were sympathetic to Iran's Islamic revolution and belonged to the Lebanese Da'wa Party. Hezbollah was responsible for several acts of anti-U.S. and anti-Israel terrorism in the 1980s and 1990s.[30] Hezbollah's attacks on Israeli forces in southern Lebanon contributed to an Israeli withdrawal in May 2000, but, despite United

Nations certification of Israel's withdrawal, Hezbollah maintained military forces along the border. Hezbollah continued to remain armed and outside Lebanese government control, despite U.N. Security Council Resolution 1559 (September 2, 2004) that required its dismantlement. In refusing to disarm, Hezbollah says it was resisting Israeli occupation of some Lebanese territory (Shib'a Farms).

Although Iran likely did not instigate Lebanese Hezbollah to provoke the July-August 2006 war, Iran has long been its major arms supplier. Hezbollah fired Iranian-supplied rockets on Israel's northern towns during the fighting. Reported Iranian shipments to Hezbollah prior to the conflict included the "Fajr" (dawn) and Khaybar series of rockets that were fired at the Israeli city of Haifa (30 miles from the border), and over 10,000 Katyusha rockets that were fired at cities within 20 miles of the Lebanese border.[31] Iran also supplied Hezbollah with an unmanned aerial vehicle (UAV), the *Mirsad*, that Hezbollah briefly flew over the Israel-Lebanon border on November 7, 2004, and April 11, 2005; at least three were shot down by Israel during the conflict. On July 14, 2006, Hezbollah apparently hit an Israeli warship with a C-802 sea-skimming missile probably provided by Iran. (See above for information on Iran's acquisition of that weapon from China.) Iran also purportedly provided advice during the conflict; about 50 Revolutionary Guards Qods Force personnel were in Lebanon (down from about 2,000 when Hezbollah was formed, according to a *Washington Post* report of April 13, 2005) when the conflict began; that number might have increased during the conflict to help Hezbollah operate the Iran-supplied weaponry.

Even though Hezbollah reduced its overt military presence in southern Lebanon in accordance with the conflict-related U.N. Security Council Resolution 1701 (July 31, 2006), Hezbollah was perceived as a victor in the war for holding out against heavy Israeli air-strikes and some ground action. Iran supported Hezbollah's demands and provided it with leverage by resupplying it, after the hostilities, with 27,000 rockets, more than double what Hezbollah had at the start of the 2006 war.[32] Among the post-war deliveries were 500 Iranian-made "Zelzal" (Earthquake) missiles with a range of 186 miles, enough to reach Tel Aviv from south Lebanon. Iran also made at least $150 million available for Hezbollah to distribute to Lebanese citizens (mostly Shiite supporters of Hezbollah) whose homes were damaged in the Israeli military campaign.[33] The State Department terrorism report for 2008, referenced above, for the first time specifies Iranian aid to Hezbollah as exceeding $200 million in 2008, and says that Iran trained over 3,000 Hezbollah fighters in Iran during the year.

Neither Israel nor the United States opposed Hezbollah's progressively increased participation in peaceful Lebanese politics. In March 2005, President George W. Bush indicated that the United States might accept Hezbollah as a legitimate political force in Lebanon if it disarms. In the Lebanese parliamentary elections of May—June 2005, Hezbollah expanded its presence in the parliament to 14 out of the 128-seat body, and it gained two cabinet seats. In mid May 2008, Hezbollah, for the first time ever, used its militia wing for domestic purposes. Its fighters took over large parts of Beirut in response to an attempt by the U.S. and Saudi-backed Lebanese government to curb Hezbollah's media and commercial operations. The success of its fighters contributed to a Qatar-brokered settlement on May 21, 2008, in which the majority coalition agreed to give Hezbollah and its allies enough seats in a new cabinet to be able to veto government decisions. Hezbollah agreed to the compromise candidate of Lebanese Army commander Michel Suleiman to become the new president. The cabinet, in which Hezbollah had one cabinet seat but its allies had seven others, giving Hezbollah its long-sought veto power, was appointed in June 2008.

Based on the strength, Hezbollah was viewed as a likely winner in June 7, 2009 parliamentary elections in Lebanon. However, its coalition won 57 seats in the elections, failing to overturn the majority of the pro-U.S. factions led by Sa'd al-Hariri, son of assassinated leader Rafiq Hariri, which won 71 seats (one more than they had previously). The defeat for Hezbollah allies set back Tehran's regional influence. As a matter of policy, the United States does not meet with any Hezbollah members, even those in the parliament or cabinet. Hezbollah is a designated FTO, but that designation bars financial transactions by the group and does not specifically ban meeting with members of the group.

Syria

Iran's support for Hezbollah is linked in many ways to its alliance with Syria. Syria is the transit point for the Iranian weapons shipments to Hezbollah and both countries see Hezbollah as leverage against Israel to achieve their regional and territorial aims. In order to preserve its links to Syria, which is one of Iran's few real allies, Iran purportedly has acted as an intermediary with North Korea to supply Syria with various forms of WMD and missile technology. Some see Israel-Syria negotiations as means to wean Syria away from its alliance with Iran by helping Syria achieve its territorial demands, particularly the Golan Heights, lost to Israel in 1967. Iran also is a major investor in the Syrian economy, which attracts very little western investment.

Central Asia and the Caspian

Iran's policy in Central Asia has thus far emphasized Iran's rights to Caspian Sea resources, particularly against Azerbaijan. That country's population, like Iran's, is mostly Shiite Muslim, but its leadership is secular. In addition, Azerbaijan is ethnically Turkic, and Iran fears that Azerbaijan nationalists might stoke separatism among Iran's large Azeri Turkic population, which demonstrated some unrest in 2006. These factors could explain why Iran has generally tilted toward Armenia, which is Christian, even though it has been at odds with Azerbaijan over territory and control of ethnic Armenians. In July 2001, Iranian warships and combat aircraft threatened a British Petroleum (BP) ship on contract to Azerbaijan out of an area of the Caspian that Iran considers its own. The United States called that action provocative, and it is engaged in border security and defense cooperation with Azerbaijan directed against Iran (and Russia). The United States successfully backed construction of the Baku-Tblisi-Ceyhan oil pipeline, intended in part to provide alternatives to Iranian oil. Along with India and Pakistan, Iran has been given observer status at the Central Asian security grouping called the Shanghai Cooperation Organization (SCO—Russia, China, Kazakhstan, Kyrgyzstan, Uzbekistan, and Tajikistan). In April 2008, Iran applied for full membership in the organization, which opposes a long-term U.S. presence in Central Asia.

Afghanistan and Pakistan[34]

Iran is viewed by U.S. officials as pursuing a multi-track strategy—attempting to help develop Afghanistan and enhance its influence there, while also building leverage against the

United States by arming anti-U.S. militant groups. Iran is particularly interested in restoring some of its traditional sway in eastern, central, and northern Afghanistan where Persian-speaking Afghans predominate.

The State Department terrorism report for 2008 again accuses the Qods Force of supplying various munitions, including 107mm rockets, to Taliban and other militants in Afghanistan; some Taliban commanders openly say they are obtaining Iranian weapons. The 2008 reports also, and for the first time, accuses Iran of training Taliban fighters in small unit tactics, small arms use, explosives, and indirect weapons fire. Among specific shipments noted by the United States: on April 17, 2007, U.S. military personnel in Afghanistan captured a shipment of Iranian weapons that purportedly was bound for Taliban fighters. On June 6, 2007 and again on September 6, 2007, NATO officers said they directly intercepted Iranian shipments of heavy arms, C4 explosives, and advanced roadside bombs (explosively forced projectiles, or EFPs, such as those found in Iraq) to Taliban fighters in Afghanistan. U.S. commanders in Afghanistan maintain that the intercepted shipments are large enough that the Iranian government would have to have known about them. U.S. and Afghan officials say the shipments continued, but are not consistent or necessarily decisive in the Afghanistan war.

These shipments and contacts have caused debate over Iran's goals because Iran long opposed the regime of the Taliban in Afghanistan on the grounds that it oppressed Shiite Muslim and other Persian-speaking minorities. Iran nearly launched a military attack against the Taliban in September 1998 after Taliban fighters captured and killed nine Iranian diplomats based in northern Afghanistan, and Iran provided military aid to the Northern Alliance factions. During the major combat phase of the post-September 11 U.S.-led war in Afghanistan, Iran offered search and rescue of any downed service-persons and the trans-shipment to Afghanistan of humanitarian assistance. Iran and U.S. diplomats were in continuous contact in forging a post-Taliban government in Afghanistan at the December 2001 "Bonn Conference." In March 2002, Iran expelled Gulbuddin Hikmatyar, an Afghan militant leader; it froze his assets in January 2005.

After 2004, Iran's influence waned somewhat as Northern Alliance figures were marginalized in Afghan politics. To rebuild that influence, Iran has funded projects in Afghanistan that total about $1.2 billion million since 2001 (close to a pledged amount in international donors conferences), mostly in neighboring Herat but also in Kabul (Shiite theological seminaries there). Iran's construction of Shiite mosques and seminaries could indicate Iran is trying to support Afghanistan's Shiite (Hazara tribe) minority, and Iran has funded several media outlets in Afghanistan catering to Shiites.

At the same time, some commanders, including CENTCOM Commander Gen. David Petraeus, have said that U.S. engagement with Iran on Afghanistan might help U.S. stabilization efforts there. Others say that working with Iran on Afghanistan might help build a broader understanding with Iran on other issues, including the nuclear issue. Some press reports say that, despite Iran's assistance to the Taliban, Defense Department planners might even be evaluating the feasibility of using an Iranian route to ship U.S. equipment into Afghanistan as an alternative to the frequently attacked route through Pakistan. Other accounts say an Iranian supply route might be used by non-U.S. partners of NATO for their missions in Afghanistan.

Perhaps in recognition of Iran's role in Afghanistan, or as part of a broader effort to build dialogue with Iran, the United States invited Iran to an international conference on

Afghanistan held in the Netherlands on March 31, 2009. Iran's representatives there had a brief side exchange there with U.S. special representative for Afghanistan and Pakistan Ambassador Richard Holbrooke. At the meeting, Iran pledged cooperation on preventing drug smuggling out of Afghanistan and in helping economically develop that country.

Pakistan

Iran's relations with Pakistan have been partly a function of events in Afghanistan. Iran had a burgeoning military cooperation with Pakistan in the early 1990s, and as noted Iran's nuclear program benefitted from the A.Q. Khan network. However, Iran-Pakistan relations became strained in the 1990s when Pakistan was supporting the Taliban in Afghanistan, which committed alleged atrocities against Shiite Afghans (Hazara tribe), and which seized control of Persian-speaking areas of Afghanistan. Currently, Iran remains suspicious that Pakistan might want to again implant the Taliban in power in Afghanistan—and Iran itself is aiding the Taliban to some extent—but Iran and Pakistan now have a broad agenda that includes a potential major gas pipeline project, discussed further below.

Al Qaeda

Iran is not a natural ally of Al Qaeda, largely because Al Qaeda is an orthodox Sunni Muslim organization. The 9/11 Commission report said several of the September 11 hijackers and other plotters, possibly with official help, might have transited Iran, but the report does not assert that the Iranian government cooperated with or knew about the plot. Another bin Laden ally, Abu Musab al-Zarqawi, killed by U.S. forces in Iraq on June 7, 2006, reportedly transited Iran after the September 11 attacks and took root in Iraq, becoming an insurgent leader there.

However, Iran might see possibilities for tactical alliance with Al Qaeda, and U.S. officials have said since January 2002 that Iran has not prosecuted or extradited senior Al Qaeda operatives (spokesman Sulayman Abu Ghaith, top operative Sayf Al Adl, and Osama bin Laden's son, Saad[35]). All have been believed to be in Iran,[36] although some U.S. officials said in January 2009 that Saad bin Laden might have left Iran and could be in Pakistan. That information was publicized a few days after the Treasury Department (on January 16, 2009) designated four Al Qaeda operatives in Iran, including Saad bin Laden (and three lesser known figures) as terrorist entities under Executive Order 13224. (U.S. officials blamed Saad bin Laden, Adl, and Abu Ghaith for the May 12, 2003, bombings in Riyadh, Saudi Arabia against four expatriate housing complexes on these operatives, saying they have been able to contact associates outside Iran.[37])

Iran asserted on July 23, 2003, that it had "in custody" senior Al Qaeda figures. However, some experts believe that hardliners in Iran might want to use Al Qaeda activists as leverage against the United States and its allies. Some say Iran might want to exchange them for a U.S. hand-over of PMOI activists under U.S. control in Iraq. Possibly attempting to show that it is an adversary and not an ally of Al Qaeda, on July 16, 2005, Iran's Intelligence Minister said that 200 Al Qaeda members are in Iranian jails.[38]

Latin America

A growing concern has been Iran's developing relations with countries and leaders in Latin America considered adversaries of the United States, particularly Cuba and Venezuela's Hugo Chavez. In February 2006, Secretary of State Rice referred to Venezuela and Cuba as "sidekicks" of Iran because of their votes in the IAEA against referring Iran to the Security Council. On January 27, 2009, Secretary of Defense Gates said Iran was trying to build influence in Latin America by expanding front companies and opening offices in countries there. On October 30, 2007, then Secretary of Homeland Security Michael Chertoff said that Iran's relationship with Venezuela is an emerging threat because it represents a "marriage" of Iran's extremist ideology with "those who have anti-American views." Chavez has visited Iran on several occasions, offering Iran additional gasoline during Iran's fuel shortages in 2007 as well as joint oil and gas projects. The two countries have established direct air links, and 400 Iranian engineers have reportedly been sent to Venezuela to work on infrastructure projects there. The State Department terrorism report for 2006 said that Cuba maintains "close relationships with other state sponsors of terrorism such as Iran." In October 2007, Uruguayan parliamentary investigators said they blocked an attempt by the government to buy arms from Iran, using a diversion through Venezuela.[39] Iran is also trying to extend its influence in Latin America by offering Bolivia $1 billion in aid and investment, according to an *Associated Press* report of November 23, 2008.

India

Iran and India have cultivated good relations with each other in order to enable each to pursue its own interests and avoid mutual conflict. The two backed similar anti-Taliban factions in Afghanistan during 1996-2001 and have a number of mutual economic and even military-to-military relationships and projects, discussed further in CRS Report RS22486, *India-Iran Relations and U.S. Interests*, by K. Alan Kronstadt and Kenneth Katzman.

One aspect of the relationship involves not only the potential building of a natural gas pipeline from Iran, through Pakistan, to India, but also the supplies of gasoline to Iran. A key supplier is Reliance Industries Ltd., which by some accounts supplies up to 40% of Iran's imports of gasoline. In December 2008, some Members of Congress expressed opposition to a decision by the Export-Import Bank to provide up to $900 million in loan guarantees to Reliance, because of its relationship with Iran. Reliance subsequently announced that it was ending new gas sales to Iran, although observers say deliveries continue. Another source of U.S. concern has been visits to India by some Iranian naval personnel.

Africa

Some Members of Congress are concerned that Iran is support radical Islamist movements in Africa. In the 111th Congress, H.Con.Res. 16 cites Hezbollah for engaging in raising funds in Africa by trafficking in "conflict diamonds." Iran also might have supplied

Islamists in Somalia with anti-aircraft and anti-tank weaponry. The possible transfer of weaponry to Hamas via Sudan was discussed above.

U.S. Policy Responses, Options, and Legislation

The February 11, 1979 fall of the Shah of Iran, a key U.S. ally, opened the long and deep rift in U.S.-Iranian relations. On November 4, 1979, radical "students" seized the U.S. Embassy in Tehran and held its diplomats hostage until minutes after President Reagan's inauguration on January 20, 1981. The United States broke relations with Iran on April 7, 1980 (just after the failed U.S. military attempt to rescue the hostages) and the two countries had only limited official contact thereafter.[40] The United States tilted toward Iraq in the 1980-1988 Iran-Iraq war, including U.S. diplomatic attempts to block conventional arms sales to Iran, providing battlefield intelligence to Iraq[41] and, during 1987-1988, direct skirmishes with Iranian naval elements in the course of U.S. efforts to protect international oil shipments in the Gulf from Iranian mines and other attacks. In one battle on April 18, 1988 ("Operation Praying Mantis"), Iran lost about a quarter of its larger naval ships in a one-day engagement with the U.S. Navy, including one frigate sunk and another badly damaged. Iran strongly disputed the U.S. assertion that the July 3, 1988, U.S. shoot-down of Iran Air Flight 655 by the *U.S.S. Vincennes* over the Persian Gulf (bound for Dubai, UAE) was an accident.

In his January 1989 inaugural speech, President George H.W. Bush laid the groundwork for a rapprochement, saying that, in relations with Iran, "goodwill begets goodwill," implying better relations if Iran helped obtain the release of U.S. hostages held by Hezbollah in Lebanon. Iran reportedly did assist in obtaining their releases, which was completed in December 1991, but no thaw followed, possibly because Iran continued to back groups opposed to the U.S.-sponsored Middle East peace process, a major U.S. priority.

Policy during the Clinton and George W. Bush Administrations

Upon taking office in 1993, the Clinton Administration moved to further isolate Iran as part of a strategy of "dual containment" of Iran and Iraq. In 1995 and 1996, the Clinton Administration and Congress added sanctions on Iran in response to growing concerns about Iran's weapons of mass destruction, its support for terrorist groups, and its efforts to subvert the Arab-Israeli peace process. The election of Khatemi in May 1997 precipitated a U.S. shift toward engagement; the Clinton Administration offered Iran official dialogue, with no substantive preconditions. In January 1998, Khatemi publicly agreed to "people-to-people" U.S.-Iran exchanges as part of his push for "dialogue of civilizations, but he ruled out direct talks. In a June 1998 speech, then Secretary of State Albright called for mutual confidence building measures that could lead to a "road map" for normalization. Encouraged by the reformist victory in Iran's March 2000 *Majles* elections, Secretary Albright, in a March 17, 2000, speech, acknowledged past U.S. meddling in Iran, announcing some minor easing of the U.S. trade ban with Iran, and promised to try to resolve outstanding claims disputes. In September 2000 U.N. "Millennium Summit" meetings, Albright and President Clinton sent a positive signal to Iran by attending Khatemi's speeches.

George W. Bush Administration Policy

The George W. Bush Administration policy priority was to prevent Iran from obtaining a nuclear weapons capability, believing that a nuclear Iran would be even more assertive in attempting to undermine U.S. objectives in the Middle East than it already is. The George W. Bush Administration undertook multi-faceted efforts to limit Iran's strategic capabilities through international diplomacy and sanctions—both international sanctions as well as sanctions enforced by its allies, outside Security Council mandate. At the same time, the Administration engaged in bilateral diplomacy with Iran on specific priority issues, such as Afghanistan and Iraq. The policy framework was supported by maintenance of a large U.S. conventional military capabilities in the Persian Gulf and through U.S. alliances with Iran's neighbors.

At times, the George W. Bush Administration considered or pursued harder options. Some Administration officials, reportedly led by Vice President Cheney, believed that policy should focus on using the leverage of possible military confrontation with Iran or on U.S. efforts to change Iran's regime.[42] Legislation in the 110th Congress indicated support for increasing U.S. sanctions and for steps to compel other foreign companies to curtail business dealings with Iran.[43]

Overview of Obama Administration Policy

Obama Administration officials say the Administration shares the goals of the previous Administration to prevent Iran from acquiring a nuclear weapons capability, as well as to curtail Iran's meddling in the affairs of its neighbors and trying to frustrate some regional U.S. initiatives. However, President Obama has said the United States would be responsive to an Iranian "unclenched fist," and that the Administration has a "new approach" that attempts consistent and broad direct diplomacy with Iran. In concert with that approach, Obama Administration officials have not indicated support for military action should Iran continue to pursue its nuclear program—although that option has not been explicitly taken "off the table" by President Obama—and they have eschewed "regime-change" in Iran to accomplish U.S. goals. As noted, U.S. officials have not pushed for new U.N. sanctions on Iran, to date. Administration officials have told U.S. partners that U.S. diplomacy with Iran is intended to complement, not supplant, joint multilateral diplomacy with Iran.

Some Obama Administration officials, including Secretary of State Clinton and Secretary of Defense Gates, have expressed public skepticism that engagement would yield changes in Iran's policies. Others, including Dennis Ross, who was named in late February 2009 as an adviser to Secretary of State Clinton for "Southwest Asia" (a formulation understood to center on Iran), believe that the United States and its partners need to present Iran with clearer incentives and clearer punishments if Iran continues to refuse cooperation on the nuclear issue. At the same time, there were several major press in June 2009, that Ambassador Ross is to be reassigned from State Department to a coordinating role at the White House on policy toward Iran, as well as Iraq and possibly also the Arab-Israeli peace process. If accurate, this could suggest White House support for the Ross approach on Iran.

Implementation of the Engagement Policy

Prior to the June 12 election in Iran, the steps to engage Iran included:

- The message to the Iranian people by President Obama on the occasion of Nowruz (Persian New Year), March 21, 2009. Experts noted particularly the President's reference to "The Islamic Republic of Iran," a formulation that appears to suggest that the United States fully accepts the Islamic revolution in Iran and is not seeking "regime change."
- The major speech to the "Muslim World" in Cairo on June 4, 2009, in which President Obama said the United States had played a role in the overthrow of Mossadeq, and said that Iran had a right to peaceful nuclear power if it complies with its responsibilities under the NPT.
- The public invitation for Iran to attend the March 31, 2009, conference on Afghanistan in the Netherlands, discussed above.
- The U.S. announcement on April 8 that it would attend all future P5+1 meetings with Iran, and suspension of seeking new P5+1 agreement on additional U.N. sanctions, pending the outcome of the Administration outreach to Iran.
- The loosening of restrictions on U.S. diplomats to meet their Iranian counterparts at international meetings, and the message to U.S. embassies abroad that they can invite Iranian diplomats to upcoming celebrations of U.S. Independence Day. (The July 4 invitations were rescinded in the aftermath of the Iran unrest.)
- On the other hand, President Obama issued a formal one year extension of the U.S. ban on trade and investment with Iran on March 15, 2009, (see "U.S. Ban on Trade and Investment with Iran," below).

Supreme Leader Khamene'i, in his Nowruz message to Iran, answered the Obama message by saying that the United States needs to take "concrete steps" to show sincerity, such as unfreezing Iranian assets, lifting U.S. sanctions, and resolving the case of the 1988 downing of an Iranian airliner (see below for further information on these issues). President Ahmadinejad's position has been that Iran is ready for dialogue with the United States if it is part of a fundamental change in the U.S. stance from what Iran sees as hostility and is "honest." However, these did not constitute the unambiguous, positive response to the overtures sought by the Administration.

Policy Implications of the Iranian Election Dispute

The dispute over the June 12 election is complicating the Administration outreach. As noted above, the Obama Administration has said its outreach policy might be slowed by the Iranian response to the unrest, and the United States and its partners now foresee a delay in the restart of nuclear talks until at least September 2009, if they restart at all. If talks do restart, it is possible that any return to discussions of new sanctions would have to wait until early 2010, in order to give new talks time to either succeed or stall. On the other hand, it is possible that Ahmadinejad and senior leaders might try to accommodate the protest movement by moderating some Iranian policies, which would in that event perhaps facilitate talks.

Enhanced U.S. Interests Section

On specific future steps toward greater engagement, the George W. Bush Administration said in late 2008 that it would leave to the Obama Administration a decision on whether to staff the U.S. interests section in Tehran with U.S. personnel, who would mostly process Iranian visas and help facilitate U.S.-Iran people-to-people contacts. The current interests section is under the auspices of the Swiss Embassy. The Obama Administration appeared inclined toward that step as well but no decision has been announced, to date, and is likely derailed by the Iranian response to the postelection protests. A potential factor in the interests section decision could be a storming of a British diplomatic facility by 50 Iranian students on December 30, 2008, protesting what they said was Britain's bias toward Israel.

Engagement Efforts during the George W. Bush Administration

Prior to 2008, the George W. Bush Administration directly engaged Iran on specific regional priority (Afghanistan and Iraq) and humanitarian issues. The United States had a dialogue with Iran on Iraq and Afghanistan from late 2001 until May 2003, when the United States broke off the talks following the May 12, 2003, terrorist bombing in Riyadh. At that time, the United States and Iran publicly acknowledged that they were conducting direct talks in Geneva on those two issues,[44] the first confirmed direct dialogue between the two countries since the 1979 revolution. The United States briefly resumed some contacts with Iran in December 2003 to coordinate U.S. aid to victims of the December 2003 earthquake in Bam, Iran, including a reported offer—rebuffed by Iran—to send a high-level delegation to Iran including Senator Elizabeth Dole and reportedly President George W. Bush's sister, Dorothy.

Prior to the July 2008 decision to have Undersecretary Burns attend the July 19, 2008 P5+1 nuclear negotiations with Iran, the George W. Bush Administration maintained it would join multilateral nuclear talks, or even potentially engage in direct bilateral talks, only if Iran suspended uranium enrichment. Some believe the Administration position was based on a view that offering to participate in a nuclear dialogue with Iran would later increase international support for sanctions by demonstrating U.S. willingness to negotiate.

The George W. Bush Administration did indicate that it considers Iran a great nation and respects its history; such themes were prominent in speeches by President George W. Bush such as at the Merchant Marine Academy on June 19, 2006, and his September 18, 2006, speech to the U.N. General Assembly. Then Secretary of State Rice said in January 2008 that the United States does not consider Iran a "permanent enemy." An amendment by then Senator Biden (adopted June 2006) to the FY2007 defense authorization bill (P.L. 109-364) supported the Administration's offer to join nuclear talks with Iran.

"Grand Bargain Concept"

The George W. Bush Administration did not offer Iran an unconditional, direct U.S.-Iran bilateral dialogue on *all* issues of U.S. concern: nuclear issues, Iranian support of militant movements, involvement in Iraq, and related issues. Some argue that the issues that divide the United States and Iran cannot be segregated, and that the key to resolving the nuclear issue is striking a "grand bargain" on all outstanding issues. The Obama Administration outreach appears to suggest a willingness to consider such a comprehensive agreement, if such agreement could be reached.

Some say the George W. Bush Administration "missed an opportunity," saying that U.S. officials rebuffed a reported comprehensive overture from Iran just before the May 12, 2003, Riyadh bombing, along the lines of a so-called "grand bargain." The *Washington Post* reported on February 14, 2007 ("*2003 Memo Says Iranian Leaders Backed Talks*"), that the Swiss Ambassador to Iran in 2003, Tim Guldimann, had informed U.S. officials of a comprehensive Iranian proposal for talks with the United States. However, State Department officials and some European diplomats based in Tehran at that time question whether that proposal represented an authoritative Iranian communication. Others argue that the offer was unrealistic because an agreement would have required Iran to abandon key tenets of its Islamic revolution.

Containment and Possible Military Action

The George W. Bush Administration consistently maintained that military action to delay or halt Iran's nuclear program was an option that was "on the table." The Obama Administration has not ruled this option out but has not indicated any inclination toward it. Although some oppose most forms of military action against Iran, others fear that military action might be the only means of preventing Iran from acquiring a working nuclear device. A U.S. ground invasion to remove Iran's regime has not, at any time, appeared to be under serious consideration in part because of the likely resistance an invasion would meet in Iran.

Proponents of U.S. air and missile strikes against suspected nuclear sites argue that military action could set back Iran's nuclear program because there are only a limited number of key targets, and these targets are known to U.S. planners and vulnerable, even those that are hardened or buried.[45] Estimates of the target set range from 400 nuclear and other WMD-related targets, to potentially a few thousand targets crucial to Iran's economy and military. At least 75 targets are underground or hardened. Those who take an expansive view of the target set argue that the United States would need to reduce Iran's potential for retaliation by striking not only nuclear facilities but also Iran's conventional military, particularly its small ships and coastal missiles.

Still others argue that there are military options that do not involve air or missile strikes. Some say that a naval embargo or related embargo is possible that could pressure Iran into reconsidering its stand on the nuclear issue. Such action was "demanded" in H.Con.Res. 362 (see below). Others say that the imposition of a "no-fly zone" over Iran might also serve that purpose. Either action could still be considered acts of war, and could escalate into hostilities.

Most U.S. allies in Europe, not to mention Russia and China, oppose military action. These states tend to agree with experts who maintain that any benefits would be temporary, and are not justified by the risks. Some believe that a U.S. strike would cause the Iranian public to rally around Iran's regime, others say a strike would provoke a new regional war. On the other hand, some European and other diplomats say that France and Britain might back or even join a military strike if Iran were to begin an all-out drive toward a nuclear weapon.

An Israeli Strike?

Israeli officials view a nuclear armed Iran as an existential threat and have repeatedly refused to rule out the possibility that Israel might strike Iran's nuclear infrastructure. Speculation about this possibility increased in March and April 2009 with statements by Israeli Prime Minister Benjamin Netanyahu to *The Atlantic* magazine stating that "You don't want a messianic apocalyptic cult controlling atomic bombs," which generated testimony in Congress by CENTCOM commander General Petraeus indicating that Israel has become so frightened by a prospect of a nuclear Iran that it might decide to launch a strike on Iran's nuclear facilities. Adding to the prospects for this scenario, in mid-June 2008, Israeli officials confirmed reports that Israel had practiced a long range strike such as that which would be required. In 2008, the George W. Bush Administration reportedly strongly discouraged an Israeli plan to conduct such a strike, and it denied Israel's requests for certain equipment useful to that operation.

Although Israeli strategists say this might be a viable option, several experts doubt that Israel has the capability to make such action sufficiently effective to justify the risks. U.S. military leaders are said by observers to believe that an Israeli strike would inevitably draw the United States into a conflict with Iran but without the degree of planning that would be needed for success.

Iranian Retaliatory Scenarios[46]

Some officials and experts warn that a U.S. military strike on Iran could provoke unconventional retaliation. At the very least, such conflict is likely to raise world oil prices significantly out of fear of an extended supply disruption. Others say such action would cause Iran to withdraw from the NPT and refuse any IAEA inspections. Other possibilities include firing missiles at Israel—and Iran's July 2008 missile tests could have been intended to demonstrate this retaliatory capability—or directing Lebanese Hezbollah or Hamas to fire rockets at Israel. Iran could also try to direct anti-U.S. militias in Iraq and Afghanistan to attack U.S. troops.

Iran has developed a strategy for unconventional warfare that partly compensates for its conventional weakness. Then CENTCOM commander Gen. John Abizaid said in March 2006 that the Revolutionary Guard Navy, through its basing and force structure, is designed to give Iran a capability to "internationalize" a crisis in the Strait of Hormuz. On January 30, 2007, his replacement at CENTCOM, Admiral William Fallon, said that "Based on my read of their military hardware acquisitions and development of tactics ... [the Iranians] are posturing themselves with the capability to attempt to deny us the ability to operate in [the Strait of Hormuz]." (General David Petraeus became CENTCOM commander in September 2008.) In July 2008 Iran again claimed it could close the Strait in a crisis but the then commander of U.S. naval forces in the Gulf, Admiral Kevin Cosgriff, backed by Joint Chiefs Chairman Mullen, said U.S. forces could quickly reopen the waterway.

Iran has nonetheless tried to demonstrate that it is a capable force in the Gulf. It has conducted at least five major military exercises since August 2006, including exercises simultaneous with U.S. exercises in the Gulf in March 2007. Iran has repeatedly stated it is capable of closing the Strait of Hormuz and would do so, if attacked. In early 2007, Iranian ships were widening their patrols, coming ever closer to key Iraqi oil platforms in the Gulf. In February 2007, Iran seized 15 British sailors that Iran said were patrolling in Iran's waters, although Britain says they were in Iraqi waters performing coalition-related searches. They were held until April 5, 2007. On January 6, 2008, the U.S. Navy reported a confrontation in which five IRGC Navy small boats approached three U.S. Navy ships to the point where they manned battle stations. The IRGC boats veered off before any shots were fired. In October 2008, Iran announced it is building several new naval bases along the southern coast, including at Jask, indicating enhanced capability to threaten the entry and exit to the Strait of Hormuz.

If there were a conflict in the Gulf, some fear that Iran might try to use large numbers of boats to attack U.S. ships, or to lay mines, in the Strait. In April 2006, Iran conducted naval maneuvers, including test firings of what Iran claims are underwater torpedoes that can avoid detection, presumably for use against U.S. ships in the Gulf, and a surface-to-sea radar-evading missile launched from helicopters or combat aircraft. U.S. military officials said the claims might be an exaggeration. The Gulf states fear that Iran will fire coastal-based cruise missiles at their oil loading or other installations across the Gulf, as happened during the Iran-Iraq war.

Containment and the Gulf Security Dialogue

The Obama Administration is continuing the efforts of its predecessor to strengthen containment of Iran by enhancing the military capabilities of U.S. regional allies. An assertive military containment component of George W. Bush Administration policy was signaled in the January 10, 2007, Iraq "troop surge" statement by President George W. Bush. In that statement, he announced that the United States was sending a second U.S. aircraft carrier group into the Gulf,[47] extending deployment of Patriot anti-missile batteries in the Gulf, reportedly in Kuwait and Qatar, and increasing intelligence sharing with the Gulf states. Secretary of Defense Gates said at the time that he saw the U.S. buildup as building leverage against Iran that could bolster diplomacy.

The U.S. Gulf deployments built on a containment strategy inaugurated in mid-2006 by the State Department, primarily the Bureau of Political-Military Affairs ("Pol-Mil"). It was termed the "Gulf Security Dialogue" (GSD), and represented an effort to revive some of the U.S.-Gulf state defense cooperation that had begun during the Clinton Administration but had since languished as the United States focused on the post-September 11 wars in Afghanistan and Iraq.

One goal of the GSD is to boost Gulf state capabilities through new arms sales to the GCC states. The emphasis of the sales is to improve Gulf state missile defense capabilities, for example by sales of the upgraded Patriot Advanced Capability-3 (PAC-3),[48] as well as to improve border and maritime security equipment through sales of combat littoral ships, radar systems, and communications gear. Several GSD-inspired sales include: PAC-3 sales to UAE and Kuwait, and Joint Direct Attack Munitions (JDAMs) to Saudi Arabia and UAE (notified to Congress in December 2007 and January 2008). A sale to UAE of the very advanced "THAAD" (Theater High Altitude Area Defense) has also been notified.

Presidential Authorities and Legislation

A decision to take military action might raise the question of presidential authorities. In the 109th Congress, H.Con.Res. 391, introduced on April 26, 2006, called on the President to not initiate military action against Iran without first obtaining authorization from Congress. A similar bill, H.Con.Res. 33, was introduced in the 110th Congress. Other bills requiring specific congressional authorization for use of force against Iran (or prohibiting U.S. funds for that purpose) include S.Res. 356, H.J.Res. 14, H.R. 3119, S.Con.Res. 13, S. 759, and H.R. 770. An amendment to H.R. 1585, the National Defense Authorization Act for FY2008, was defeated 136 to 288. A provision that sought to bar the Administration from taking military action against Iran without congressional authorization was taken out of an early draft of an FY2007 supplemental appropriation (H.R. 1591) to fund additional costs for Iraq and Afghanistan combat (vetoed on May 1, 2007). Other provisions, including requiring briefings to Congress about military contingency planning related to Iran's nuclear program, is in the House-passed FY2009 defense authorization bill (H.R. 5658). In the 111th Congress, H.Con.Res. 94 calls for the United States to negotiate an "Incidents at Sea" agreement with Iran.

Regime Change

A major early feature of George W. Bush Administration policy—promotion of "regime change"—receded in the latter stages of the Administration. The Obama Administration has clearly distanced itself from the prior Administration's attraction to this option, for example by explicitly referring to Iran by its formal name taken after the Islamic revolution – "the Islamic Republic of Iran." There are no indications, to date, that the Obama Administration sees the postelection protests in Iran as providing the United States an opportunity to change the regime.

There has been some support in the United States for regime change since the 1979 Islamic revolution; the United States provided some funding to anti-regime groups, mainly promonarchists, during the 1980s.[49] The George W. Bush Administration's belief in this option became apparent after the September 11, 2001, attacks, when President George W. Bush described Iran as part of an "axis of evil" in his January 2002 State of the Union message. President George W. Bush's second inaugural address (January 20, 2005) and his State of the Union messages of February 2, 2005, and January 31, 2006, suggested a clear preference for a change of regime by stating, in the latter speech, that "... our nation hopes one day to be the closest of friends with a free and democratic Iran." Indications of affinity for this option include increased public criticism of the regime's human rights record as well as the funding of Iranian pro-democracy activists. However, the George W. Bush Administration shifted away from this option as a strategy employing multilateral sanctions and diplomacy took form in 2006, in part because U.S. partners believe regime change policies harm diplomacy.

Although it was clearly hoping for opportunities to change the regime, the George W. Bush Administration said that the democracy promotion programs discussed below were intended to promote political evolution in Iran and change regime behavior, not to overthrow the regime. A few accounts, such as "*Preparing the Battlefield*" by Seymour Hersh in the New Yorker (July 7 and 14, 2008) say that President George W. Bush authorized U.S. covert operations to destabilize the regime,[50] involving assistance to some of the ethnic-based armed groups discussed above. CRS has no way to confirm assertions in the Hersh article that up to

$400 million was appropriated and/or used to aid the groups mentioned. In January 2009, Iran tried four Iranians on charges of trying to overthrow the government with U.S. support.

Democracy Promotion Efforts

The George W. Bush Administration's efforts to promote democracy in Iran began in FY2004 and were supported by many in Congress. Clear congressional sentiment in favor of this effort came in the 109th Congress with enactment of the Iran Freedom Support Act (H.R. 6198, P.L. 109-293, signed September 30, 2006, which authorized funds (no specific dollar amount) for Iran democracy promotion and modified the Iran Sanctions Act.[51]

Many question the prospects of U.S.-led Iran regime change through democracy promotion because of the weakness of opposition groups. Providing overt or covert support to anti-regime organizations, in the view of many experts, would not make them materially more viable or attractive to Iranians. The regime purportedly also conducts extensive regime surveillance of democracy activists or other internal dissidents. Iran has been arresting civil society activists by alleging they are accepting the U.S. democracy promotion funds, while others have refused to participate in U.S.-funded programs, fearing arrest. The highest profile such arrest came in May 2007 – Iranian-American scholar Haleh Esfandiari, of the Woodrow Wilson Center in Washington, DC, was imprisoned for several months. [52]

The State Department has been the implementer of U.S. democracy promotion programs. The Department has used funds in appropriations (see below) to support pro-democracy programs run by 26 organizations based in the United States in Europe; the Department refuses to name grantees for security reasons. In 2006, the George W. Bush Administration also began increasing the presence of Persian-speaking U.S. diplomats in U.S. diplomatic missions around Iran, in part to help identify and facilitate Iranian participate in U.S. democracy-promotion programs. The Iran unit at the U.S. consulate in Dubai has been enlarged significantly into a "regional presence" office, and new "Iran-watcher" positions have been added to U.S. diplomatic facilities in Baku, Azerbaijan; Istanbul, Turkey; Frankfurt, Germany; London; and Ashkabad, Turkmenistan, all of which have large expatriate Iranian populations and/or proximity to Iran.[53] An enlarged (eight person) "Office of Iran Affairs" has been formed at State Department, and it is reportedly engaged in contacts with U.S.-based exile groups such as those discussed earlier. Iran asserts that funding democracy promotion represents a violation of the 1981 "Algiers Accords" that settled the Iran hostage crisis and provide for non-interference in each others' internal affairs.

The Obama Administration has not, to date, announced a discontinuation of the State Department efforts, but has rather appeared to shift the emphasis to the people-to-people exchange component of these programs. To date the State Department has sponsored exchanges with about 150 Iranian academics, professionals, athletes, artists, and doctors. Another part of the effort is broadcasting to Iran. As noted below, the Voice of America and Radio Free Europe/Radio Liberty have been expanding broadcasts to Iran of information about Iran and about the United States. The Department has also formed a Persian-language website.

Table 9. Iran Democracy Promotion Funding

FY2004	Foreign operations appropriation (P.L. 108-199) earmarked $1.5 million for "educational, humanitarian and non-governmental organizations and individuals inside Iran to support the advancement of democracy and human rights in Iran." The State Department Bureau of Democracy and Labor (DRL) gave $1 million to a unit of Yale University, and $500,000 to National Endowment for Democracy.
FY2005	$3 million from FY2005 foreign aid appropriation (P.L. 108-447) for democracy promotion. Priority areas: political party development, media, labor rights, civil society promotion, and human rights.
FY2006	$11.15 for democracy promotion from regular FY2006 foreign aid appropriation (P.L. 109-102). $4.15 million administered by DRL and $7 million for the Bureau of Near Eastern Affairs.
FY2006 supp.	Total of $66.1 million (of $75 million requested) from FY2006 supplemental (P.L. 109-234): $20 million for democracy promotion ($5 million above request); $5 million for public diplomacy directed at the Iranian population (amount requested); $5 million for cultural exchanges (amount requested); and $36.1 million for Voice of America-TV and "Radio *Farda*" broadcasting ($13.9 million less than request). Of all FY2006 funds, the State Department said on June 4, 2007 that $16.05 million was obligated for democracy promotion programs, as was $1.77 million for public diplomacy and $2.22 million for cultural exchanges (bringing Iranian professionals and language teachers to the United States). Broadcasting funds provided through the Broadcasting Board of Governors; began under Radio Free Europe/Radio Liberty (RFE/RL), in partnership with the VOA, in October 1998. *Farda* ("Tomorrow" in Farsi) received $14.7 million of FY2006 funds; now broadcasts 24 hours/day. VOA Persian services (radio and TV) combined cost about $10 million per year. VOA-TV began on July 3, 2003, and now is broadcasting to Iran 12 hours a day. (Farda began when Congress funded it at $4 million in the FY1998 Commerce/State/Justice appropriation, P.L. 105-119. It was to be called Radio Free Iran but was never formally given that name by RFE/RL.)
FY2007	FY2007 continuing resolution provided $6.55 million for Iran (and Syria) to be administered through DRL. $3.04 million was used for Iran. No funds were requested.
FY2008	$60 million (of $75 million requested) is contained in Consolidated Appropriation (H.R. 2764, P.L. 110- 161), of which $21.6 million is ESF for pro-democracy programs, including non-violent efforts to oppose Iran's meddling in other countries. $7.9 million is "Development Funds" for use by DRL. Appropriation also fully funds additional $33.6 million requested for Iran broadcasting: $20 million for VOA Persian service; and $8.1 million for Radio Farda; and $5.5 million for exchanges with Iran.
FY2009	Request was for $65 million in ESF "to support the aspirations of the Iranian people for a democratic and open society by promoting civil society, civic participation, media freedom, and freedom of information." H.R. 1105 (P.L. 111-8) provides $15 million for democracy promotion programs in Iran and several other countries.
FY2010	No specific democracy promotion request, but some funds (out of $40 million requested for Near East democracy programs) likely to fund continued people-to-people exchanges with Iran.

Funding

As shown below, $67 million has been appropriated for Iran democracy promotion ($19.6 million through DRL and $48.6 million through the Bureau of Near Eastern Affairs/USAID). Additional funds, discussed in the chart below, have been appropriated for cultural exchanges, public diplomacy, and broadcasting to Iran. The Obama Administration requested funds for Near East regional democracy programs in its FY2010 budget request, but no specific request for funds for Iran were delineated. This could be an indication that the new Administration views this effort as inconsistent with its belief in dialogue with Iran. No U.S. assistance has been provided to exile-run stations. (The conference report on the FY2006 regular foreign aid appropriations, P.L. 109-102, stated the sense of Congress that such support should be considered.)

Further International and Multilateral Sanctions

The Obama Administration, although promoting the concept of sustained dialogue, says it will return to efforts to increase international pressure on Iran if efforts at direct diplomacy do not yield results by the end of 2009, although that deadline is now likely to slip, according to Administration officials. The following represent sanctions that the Security Council might impose, along with some discussion of key positions expressed by some Security Council or other nations on those ideas. U.S. officials have said that these or other additional sanctions might also be considered by a "coalition" of countries, outside Security Council authorization. On the other hand, some U.S. allies, such as Germany, oppose sanctions outside Council action on the grounds that doing so would undermine the Security Council process.[54] Among the further U.N. or multilateral sanctions widely discussed (and some of these ideas are appearing in U.S. legislation to increase U.S. sanctions on Iran) are:

- *Mandating Reductions in Diplomatic Exchanges with Iran or Prohibiting Travel by Iranian Officials.* As noted above, Resolution 1803 imposes a ban on travel by some named Iranian officials. One option is to further expand that list of Iranian officials. A further option is to limit sports or cultural exchanges with Iran, such as Iran's participation in the World Cup soccer tournament. However, many experts oppose using sporting events to accomplish political goals.
- *Banning International Flights to and from Iran.* Bans on flights to and from Libya were imposed on that country in response to the finding that its agents were responsible for the December 21, 1988, bombing of Pan Am 103 (now lifted). There are no indications that a passenger aircraft flight ban is under consideration among the P5+1. As noted above, inspections of Iranian international cargo flights and shipping is authorized in Resolution 1803.
- *A Ban on Exports to Iran of Refined Oil Products or of Other Products.* Some members of the U.N. Security Council oppose this sanction as likely to halt prospects for a diplomatic solution to Iran's nuclear program, and the sanction does not appear to be under Security Council consideration. A gasoline exports ban would almost certainly hurt Iran's economy because Iran does not refine enough gasoline to meet demand and must import it, although some experts believe Iran would be able to circumvent this sanction by offering premium prices to suppliers. Some countries that supply gasoline to Iran, such as those listed in Table 3, might oppose this sanction.
- *Financial and Trade Sanctions, Such as a Freeze on Iran's Financial Assets Abroad.* Existing U.N. resolutions do not freeze all Iranian assets abroad, and such a broad freeze does not appear to be under Security Council consideration at this time. A

future resolution could add more Iranian banks to those under sanction, or even entirely ban transactions with any Iranian banks. Fearing that possibility, Iran moved $75 billion out of European banks in May 2008.

- *Limiting Lending to Iran by Banks or International Financial Institutions.* Another option is to ban lending to Iran by international financial institutions, or to mandate a reduction of official credit guarantees. British Prime Minister Brown indicated British support a limitation of official credits on November 12, 2007. As discussed below, EU countries and their banks have begun taking these steps, even without a specific U.N. mandate.
- *Banning Worldwide Investment in Iran's Energy Sector.* This option would represent an "internationalization" of the U.S. "Iran Sanctions Act," which is discussed in CRS Report RS20871, *The Iran Sanctions Act (ISA)*, by Kenneth Katzman. On November 12, 2007 comments, British Prime Minister Brown expressed support for a worldwide financing of energy projects in Iran as a means of cutting off energy development in Iran. A version of this option would prevent companies of U.N. member states from shipping to Iran parts or technology needed to construct oil refineries or related installations.
- *Banning Insurance for Iranian Shipping.* One option, reportedly under consideration by the P5+1, is to ban the provision of insurance for Iran's tanker fleet. Shipments of Iranian oil require insurance against losses from military action, accidents, or other causes. A broad ban on provision of such insurance could make it difficult for Iran to Islamic Republic of Iran Shipping Lines (IRISL) to operate and force Iran to rely on more expensive shipping options. Iran said in September 2008 that it would have ways to circumvent the effect of this sanction if it is imposed. (The United States has imposed sanctions on IRISL, as noted in the table below.)
- *Imposing a Worldwide Ban on Sales of Arms to Iran.* Resolution 1747 called for—but did not require—U.N. member states to exercise restraint in selling arms to Iran. A future resolution might mandate an arms sales ban. Another option under discussion is to eliminate the Resolution 1737 exemption from sanctions for the Bushehr nuclear reactor project.
- *Imposing an International Ban on Purchases of Iranian Oil or Other Trade.* This is widely considered the most sweeping of sanctions that might be imposed, and would be unlikely to be considered in the Security Council unless Iran was found actively developing an actual nuclear weapon. Virtually all U.S. allies conduct extensive trade with Iran, and would oppose sanctions on trade in civilian goods with Iran. A ban on oil purchases from Iran is unlikely to be imposed because of the potential to return world oil prices to the high levels of the summer of 2008.

European/Japanese/Other Foreign Country Policy on Sanctions and Trade Agreements

Most U.S. allies still favor incentives—not just economic or political punishments—as tools to change Iran's behavior. In this, U.S. allies might identify with the Obama Administration approach more so than the George W. Bush Administration approach that was perceived as primarily punitive. During 1992-1997, when the United States was tightening its own sanctions against Iran, the European Union (EU) countries maintained a policy of "critical dialogue" with Iran, and the EU and Japan refused to join the 1995 U.S. trade and investment ban on Iran. The European dialogue with Iran was suspended in April 1997 in response to the German terrorism trial ("Mykonos trial") that found high-level Iranian involvement in killing Iranian dissidents in Germany, but resumed in May 1998 during Khatemi's presidency.

With Iran defiant on nuclear issues, the European countries, Japan, and other countries moved closer to the U.S. position since 2005. The EU is no longer negotiating new trade agreements and other economic interaction with Iran, but rather it has begun to implement some sanctions that exceed those mandated in Security Council resolutions. For example, several EU countries are discouraging their companies from making any new investments in or soliciting any new business with Iran. In addition, several EU countries report that civilian trade with Iran is down because Iran's defiance on the nuclear issue is introducing more perceived risk to trading with Iran. As noted above, some EU countries say they have reduced credit guarantee exposure to Iran since Resolution 1737 was passed, as shown in the table above. Previously, the EU countries and their banks maintained that financing for purely civilian goods is not banned by any U.N. resolution and that exporters of such goods should not be penalized.

Negotiations with Iran on a "Trade and Cooperation Agreement" (TCA) are not currently being held; such an agreement would have lowered the tariffs or increased quotas for Iranian exports to the EU countries. [55] Similarly, there is insufficient international support to grant Iran membership in the World Trade Organization (WTO) until there is progress on the nuclear issue. Iran first attempted to apply to join the WTO in July 1996. On 22 occasions after that, representatives of the Clinton and then the George W. Bush Administration blocked Iran from applying (applications must be by consensus of the 148 members). As discussed above, as part of an effort to assist the EU-3 nuclear talks with Iran, at a WTO meeting in May 2005, no opposition to Iran's application was registered, and Iran formally began accession talks.

In the 1990s, European and Japanese creditors—over U.S. objections—rescheduled about $16 billion in Iranian debt. These countries (governments and private creditors) rescheduled the debt bilaterally, in spite of Paris Club rules that call for multilateral rescheduling. Iran's improved external debt led most European export credit agencies to restore insurance cover for exports to Iran. In July 2002, Iran tapped international capital markets for the first time since the Islamic revolution, selling $500 million in bonds to European banks.

World Bank Loans

The EU and Japan appear to have made new international lending to Iran contingent on Iran's response to international nuclear demands. This represents a narrowing of past differences between the United States and its allies on this issue. Acting under provisions of successive foreign aid laws (which require the United States to vote against international loans to countries named by the United States as sponsors of international terrorism), in 1993 the United States voted its 16.5% share of the World Bank against loans to Iran of $460 million for electricity, health, and irrigation projects, but the loans were approved. To block that lending, the FY1994-FY1996 foreign aid appropriations (P.L. 103-87, P.L. 103-306, and P.L. 104-107) cut the amount appropriated for the U.S. contribution to the Bank by the amount of those loans. The legislation contributed to a temporary halt in new Bank lending to Iran.

During 1999-2005, Iran's moderating image had led the World Bank to consider new loans over U.S. opposition. In May 2000, the United States' allies outvoted the United States to approve $232 million in loans for health and sewage projects. During April 2003-May 2005, a total of $725 million in loans were approved for environmental management, housing reform, water and sanitation projects, and land management projects, in addition to $400 million in loans for earthquake relief.

U.S. Sanctions

Any additional international or U.S. sanctions would add to the wide range of U.S. sanctions in place since the November 4, 1979, seizure of the U.S. hostages in Tehran.[56] Some experts believe that, even before U.S. allies had begun to impose some sanctions on Iran, U.S. sanctions alone were slowing Iran's economy.[57] However, a report on U.S. sanctions by the Government Accountability Office (GAO), published December 2007 (*GAO-08-58: Iran Sanctions: Impact in Furthering U.S. Objectives Is Unclear and Should Be Reviewed*) found that the extent of the impact on Iran is "difficult to determine." The GAO report said that, despite the U.S. sanctions, Iran's global trade has continued to expand from 1987 (when sanctions first began to be imposed) to 2006, and that Iran had signed $20 billion in energy investment deals with foreign firms.

Section 7043 of P.L. 111-8 (March 11, 2009), the FY09 omnibus appropriation, requires, within 180 days, an Administration report on U.S. sanctions, including which companies are believed to be violators, and what the Administration is doing to enforce sanctions. The provision appears to apply primarily to the Iran Sanctions Act. Possible additional legislation in the 111th Congress focuses on denying U.S. government contracts to telecommunications firms that sell to Iran equipment that Iran could use to track internet use by its citizens, which Iran has done to crack down on protesters against the June 2009 election.

Terrorism/Foreign Aid Sanctions

In January 1984, following the October 1983 bombing of the U.S. Marine barracks in Lebanon (believed perpetrated by Hezbollah) Iran was added to the "terrorism list." The list was established by Section 6(j) of the Export Administration Act of 1979, sanctioning countries determined to have provided repeated support for acts of international terrorism.

- The terrorism list designation restricts sales of U.S. dual use items (Export Administration Act, as continued by executive order), and, under other laws, bans direct U.S. financial assistance (Foreign Assistance Act, FAA) and arms sales (Arms Export Control Act), and requires the United States to vote to oppose multilateral lending to the designated countries (Anti-Terrorism and Effective Death Penalty Act of 1996, P.L. 104-132). Waivers are provided under these laws, but successive foreign aid appropriations laws since the late 1980s ban direct assistance to Iran (loans, credits, insurance, Eximbank credits) without providing for a waiver.
- Section 307 of the FAA (added in 1985) names Iran as unable to benefit from U.S. contributions to international organizations, and require proportionate cuts if these institutions work in Iran. No waiver is provided for.

- The Anti-Terrorism and Effective Death Penalty Act also requires the President to withhold U.S. foreign assistance to any country that provides to a terrorism list country foreign assistance or arms. Waivers are provided.
- Executive Order 13324 (September 23, 2001) authorizes the President to freeze the assets of and transactions with entities determined to be supporting international terrorism. Iranian entities named and sanctioned under this order are in the tables at the end of this paper.

U.S. sanctions laws do not bar disaster aid and the United States donated $125,000, through relief agencies, to help victims of two earthquakes in Iran (February and May 1997), and another $350,000 worth of aid to the victims of a June 22, 2002 earthquake. (The World Bank provided some earthquake related lending as well.) The United States provided $5.7 million in assistance (out of total governmental pledges of about $32 million, of which $17 million have been remitted) to the victims of the December 2003 earthquake in Bam, Iran, which killed as many as 40,000 people and destroyed 90% of Bam's buildings. The United States military flew in 68,000 kilograms of supplies to Bam.

Proliferation Sanctions

Iran is prevented from receiving advanced technology from the United States under relevant and Iran-specific anti-proliferation laws[58] and by Executive Order 13382 (June 28, 2005). The Iran-Iraq Arms Nonproliferation Act (P.L. 102-484) requires denial of license applications for exports to Iran of dual use items, and imposes sanctions on foreign countries that transfer to Iran "destabilizing numbers and types of conventional weapons," as well as WMD technology. The Iran Nonproliferation Act (P.L. 106-178, now called the Iran-Syria-North Korea NonProliferation Act) authorizes sanctions on foreign entities that assist Iran's WMD programs. It bans U.S. extraordinary payments to the Russian Aviation and Space Agency in connection with the international space station unless the President can certify that the agency or entities under its control had not transferred any WMD or missile technology to Iran within the year prior.[59] (A Continuing Resolution for FY2009, which funded the U.S. government through March 2009, waived this law to allow NASA to continue to use Russian vehicles to access the International Space Station.)

Executive Order 13382 allows the President to block the assets of proliferators of weapons of mass destruction (WMD) and their supporters under the authority granted by the International Emergency Economic Powers Act (IEEPA, 50 U.S.C. 1701 et seq.), the National Emergencies Act (50 U.S.C. 1601 et seq.), and Section 301 of Title 3, United States Code.

The George W. Bush Administration decided to impose sanctions for violations, and it sanctioned numerous entities as discussed below. The Obama Administration sanctioned several entities on February 2, 2009, suggesting it is continuing that policy. Iranian entities designated under these laws and orders are listed in the tables at the end.

Despite these efforts, Iran has used loopholes and other devices, such as front companies, to elude U.S. and international sanctions. Some of these efforts focus on countries perceived as having lax enforcement of export control laws, such as UAE and Malaysia. In some cases, Iran has been able, according to some reports, to obtain sophisticated technology even from U.S. firms.[60]

In addition, successive foreign aid appropriations punish the Russian Federation for assisting Iran by withholding 60% of any U.S. assistance to the Russian Federation unless it terminates technical assistance to Iran's nuclear and ballistic missiles programs.

Targeted Financial Measures by Treasury Department

U.S. officials, particularly Undersecretary of the Treasury Stuart Levey (who has remained in the Obama Administration), say the United States is having substantial success in separate unilateral efforts ("targeted financial measures") to persuade European governments and companies to stop financing commerce with Iran on the grounds that doing so entails financial risk and furthers terrorism and proliferation. Treasury and State Departments officials, in April 17, 2008, testimony before the House Foreign Affairs Committee, said they had persuaded at least 40 banks not to provide financing for exports to Iran or to process dollar transactions for Iranian banks. Among those that have pulled out of Iran are UBS and Credit Suisse (Switzerland), HSBC (Britain), Germany's Commerzbank A.G and Deutsche Bank AG. U.S. pressure has reportedly convinced Kuwaiti banks to stop transactions with Iranian accounts,[61] and South Korean banks are considering doing the same. The International Monetary Fund and other sources report that these measures are making it more difficult to fund energy industry and other projects in Iran and for importers/exporters to conduct trade in expensive itesm.

Some of these moves by European banks have come about by U.S. pressure. In 2004, the Treasury Department fined UBS $100 million for the unauthorized movement of U.S. dollars to Iran and other sanctioned countries, and in December 2005, the Treasury Department fined Dutch bank ABN Amro $80 million for failing to fully report the processing of financial transactions involving Iran's Bank Melli (and another bank partially owned by Libya).

In action intended to cut Iran off from the U.S. banking system, on September 6, 2006, the Treasury Department barred U.S. banks from handling any indirect transactions ("U-turn transactions, meaning transactions with non-Iranian foreign banks that are handling transactions on behalf of an Iranian bank) with Iran's Bank Saderat (see above), which the Administration accuses of providing funds to Hezbollah.[62] Bank Sepah is subject to asset freezes and transactions limitations as a result of Resolution 1737 and 1747. The Treasury Department extended that U-Turn restriction to all Iranian banks on November 6, 2008.

Thus far, the Treasury Department has not designated any bank as a "money laundering entity" for Iran-related transactions (under Section 311 of the USA Patriot Act), although some say that step has been threatened at times. Nor has Treasury imposed any specific sanctions against Bank Markazi (Central Bank) which, according to a February 25, 2008, *Wall Street Journal* story, is helping other Iranian banks circumvent the U.S. and U.N. banking pressure. However, the European countries reportedly oppose such a sanction as an extreme step with potential humanitarian consequences, for example by preventing Iran from keeping its currency stable. S. 3445, a Senate bill in the 110th Congress, and a counterpart passed by the House on September 26, 2008, H.R. 7112, call for this sanction.

In enforcing U.S. sanctions, on December 17, 2008, the U.S. Attorney for the Southern District of New York filed a civil action seeking to seize the assets of the Assa Company, a U.K-chartered entity. Assa allegedly was maintaining the interests of Bank Melli in an office building in New York City. An Iranian foundation, the Alavi Foundation, allegedly is an investor in the building.

U.S. Ban on Trade and Investment with Iran

On May 6, 1995, President Clinton issued Executive Order 12959 banning U.S. trade and investment in Iran.[63] This followed an earlier March 1995 executive order barring U.S. investment in Iran's energy sector. The trade ban was intended to blunt criticism that U.S. trade with Iran made U.S. appeals for multilateral containment of Iran less credible. Each March since 1995 (and most recently on March 11, 2009), the U.S. Administration has renewed a declaration of a state of emergency that triggered the investment ban. Some modifications to the trade ban since 1999 account for the trade between the United States and Iran. H.R. 1400, S. 970, S. 3445, and H.R. 7112, see below, would have reimposed many of the restrictions.

The following conditions and modifications, as administered by the Office of Foreign Assets Control (OFAC) of the Treasury Department, apply:

- Some goods related to the safe operation of civilian aircraft may be licensed for export to Iran, and as recently as September 2006, the George W. Bush Administration, in the interests of safe operations of civilian aircraft, permitted a sale by General Electric of Airbus engine spare parts to be installed on several Iran Air passenger aircraft (by European airline contractors).
- U.S. firms may not negotiate with Iran or to trade Iranian oil overseas. The trade ban permits U.S. companies to apply for licenses to conduct "swaps" of Caspian Sea oil with Iran. However, a Mobil Corporation application to do so was denied in April 1999.
- Since April 1999, commercial sales of food and medical products to Iran have been allowed, on a case-by-case basis and subject to OFAC licensing. According to OFAC in April 2007, licenses for exports of medicines to treat HIV and leukemia are routinely expedited for sale to Iran, and license applications are viewed favorably for business school exchanges, earthquake safety seminars, plant and animal conservation, and medical training in Iran. Private letters of credit can be used to finance approved transactions, but no U.S. government credit guarantees are available, and U.S. exporters are not permitted to deal directly with Iranian banks. The FY2001 agriculture appropriations law (P.L. 106-387) contained a provision banning the use of official credit guarantees for food and medical sales to Iran and other countries on the U.S. terrorism list, except Cuba, although allowing for a presidential waiver to permit such credit guarantees. Neither the Clinton Administration nor the George W. Bush Administration provided the credit guarantees.
- In April 2000, the trade ban was further eased to allow U.S. importation of Iranian nuts, dried fruits, carpets, and caviar. The United States was the largest market for Iranian carpets before the 1979 revolution, but U.S. anti-dumping tariffs imposed on Iranian products in 1986 dampened of many Iranian products. The tariff on Iranian carpets is now about 3%-6%, and the duty on Iranian caviar is about 15%. In December 2004, U.S. sanctions were further modified to allow Americans to freely engage in ordinary publishing activities with entities in Iran (and Cuba and Sudan). As of mid-2007, the product most imported from Iran by U.S. importers is

pomegranate juice concentrate. In the 110th Congress, H.R. 1400, S. 970, S. 3445, and H.R. 7112 would have re-imposed the full import ban.

Application to Foreign Subsidiaries of U.S. Firms

The U.S. trade ban does not bar subsidiaries of U.S. firms from dealing with Iran, as long as the subsidiary has no operational relationship to the parent company. Among major subsidiaries that have traded with Iran are the following:

- Halliburton. On January 11, 2005, Iran said it had contracted with U.S. company Halliburton, and an Iranian company, Oriental Kish, to drill for gas in Phases 9 and 10 of South Pars. Halliburton reportedly provided $30 million to $35 million worth of services per year through Oriental Kish, leaving unclear whether Halliburton would be considered in violation of the U.S. trade and investment ban or the Iran Sanctions Act (ISA)[64]—because the deals involved a subsidiary of Halliburton (Cayman Islands-registered Halliburton Products and Service, Ltd, based in Dubai). On April 10, 2007, Halliburton announced that its subsidiaries had, as promised in January 2005, no longer operating in Iran.
- General Electric (GE). The firm announced in February 2005 that it would seek no new business in Iran, and it reportedly wound down pre-existing contracts by July 2008. GE was selling Iran equipment and services for hydroelectric, oil and gas services, and medical diagnostic projects through Italian, Canadian, and French subsidiaries.
- Foreign subsidiaries of several other U.S. energy equipment firms are apparently still in the Iranian market. These include Foster Wheeler, Natco Group, Overseas Shipholding Group,[65] UOP (a Honeywell subsidiary), Itron, Fluor, Flowserve, Parker Drilling, Vantage Energy Services, Weatherford, and a few others.
- An Irish subsidiary of the Coca Cola company provides syrup for the U.S.-brand soft drink to an Iranian distributor, Khoshgovar. Local versions of both Coke and of Pepsi (with Iranian-made syrups) are also marketed in Iran by distributors who licensed the recipes for those soft drinks before the Islamic revolution and before the trade ban was imposed on Iran.

In the 110th Congress, S. 970, S. 3227, S. 3445, and three House-passed bills (H.R. 1400, H.R. 7112, and H.R. 957)—would have applied sanctions to the parent companies of U.S. subsidiaries if those subsidiaries are directed or formed to trade with Iran.

The Iran Sanctions Act (ISA)

The Iran Sanctions Act penalizes foreign (or U.S.) investment of more than $20 million in one year in Iran's energy sector.[66] No projects have actually been sanctioned under ISA, and numerous investment agreements with Iran since its enactment have helped Iran slow deterioration of its energy export sector. On the other hand, most recent deals are with Asian energy firms that might lack the technological sophistication and effectiveness of European companies, which appear to have been dissuaded from new investments in Iran by ISA and by the policies of their governments toward Iran. (A chart illustrating the list of publicly announced investments is in CRS Report RS20871, *The Iran Sanctions Act (ISA)*, by Kenneth Katzman).

Recent legislation has sought to expand the authorities of ISA. In the 110th Congress, H.R. 1400, passed by the House on September 25, 2007, would remove the Administration's ability to waive application of sanctions under ISA. The George W. Bush Administration opposed that provision on the grounds that requiring sanctions on allied companies would divide the United States and its allies on Iran policy. However, H.R. 1400 did not impose on the Administration a time limit to determine whether a project is sanctionable. Several bills in the 110th Congress (H.R. 1400, S. 970, S. 3227, S. 3445, H.R. 7112, and H.R. 957) would have expanded the definitions of sanctionable activities and entities to include official credit guarantee agencies, such as France's COFACE and Germany's Hermes, and both bills would also clearly apply ISA sanctions to pipeline and liquified natural gas (LNG) projects. Another bill in the 110th Congress, H.R. 2880, would apply the Iran Sanctions Act (see below) to entities that sell gasoline to Iran. H.Con.Res. 362 and S.Res. 580 called for (but did not mandate) that sanction.

In the 111th Congress, several bills pick up on many of these ideas. The theme of the recently introduced bills, including H.R. 2194 and S. 908, focus on amending ISA to make sanctionable sales to Iran of gasoline, shipping of such gasoline, and sales to Iran of equipment or services to help Iran build or expand its oil refineries. These bills would, in addition, expand the menu of sanctions available for the President to impose, in order to enact penalties against violating firms. The new available sanctions would include (1) prohibiting transactions in foreign exchange by the sanctioned firm; (2) prohibiting any financial transactions on behalf of the sanctioned firm; or (3) prohibiting any acquisitions or ownership of U.S. property by the sanctioned entity.

In the aftermath of Iran's crackdown on post-election protests, the House Appropriations Committee marked up a version of a FY2010 foreign aid appropriation that would deny Eximbank credits to any firm that sells gasoline or performs gasoline production projects in Iran.

Some Members who have introduced these bills have said that such legislation might appear to conflict with President Obama's diplomatic outreach to Iran, although others believe such bills might strengthen that approach by demonstrating to Iran that there are substantial downsides to rebuffing the U.S. overtures.

Divestment

A growing trend not only in Congress but in several states is to require or call for or require divestment of shares of firms that have invested in Iran's energy sector (at the same levels considered sanctionable under the Iran Sanctions Act).[67] Legislation in the 110th Congress, H.R. 1400, did not require divestment, but requires a presidential report on firms that have invested in Iran's energy sector. Another bill, H.R. 1357, required government pension funds to divest of shares in firms that have made ISA-sanctionable investments in Iran's energy sector and bar government and private pension funds from future investments in such firms. Two other bills, H.R. 2347 (passed by the House on July 31, 2007) and S. 1430, would protect mutual fund and other investment companies from shareholder action for any losses that would occur from divesting in firms that have investing in Iran's energy sector. In the 111th Congress, H.R. 1327, a bill similar to H.R. 2347 of the 110th Congress, was reported by the Financial Services Committee on April 28, 2009.

Counter-Narcotics

In February 1987, Iran was first designated as a state that failed to cooperate with U.S. anti-drug efforts or take adequate steps to control narcotics production or trafficking. U.S. and U.N. Drug Control Program (UNDCP) assessments of drug production in Iran prompted the Clinton Administration, on December 7, 1998, to remove Iran from the U.S. list of major drug producing countries. This exempts Iran from the annual certification process that kept drug-related U.S. sanctions in place on Iran. According to several governments, over the past few years Iran has augmented security on its border with Afghanistan in part to prevent the flow of narcotics from that country into Iran. Britain has sold Iran some night vision equipment and body armor for the counter-narcotics fight.

Travel-Related Guidance

Use of U.S. passports for travel to Iran is permitted. Iranians entering the United States are required to be fingerprinted, and Iran has imposed reciprocal requirements. In May 2007, the State Department increased its warnings about U.S. travel to Iran, based largely on the arrests of the dual Iranian-American nationals discussed earlier.

Table 10. Entities Sanctioned under U.N. Resolutions and U.S. Laws and Executive Orders (Persons listed are identified by the positions they held when designated; some have since changed.)

Entities Named for Sanctions Under Resolution 1737		
Atomic Energy Organization of Iran (AEIO)	Mesbah Energy Company (Arak supplier)	Kalaye Electric (Natanz supplier)
Pars Trash Company (centrifuge program)	Farayand Technique (centrifuge program)	Defense Industries Organization (DIO)
7th of Tir (DIO subordinate)	Shahid Hemmat Industrial Group (SHIG)—missile program	Shahid Bagheri Industrial Group (SBIG) missile program
Fajr Industrial Group (missile program)	Mohammad Qanadi, AEIO Vice President	Behman Asgarpour (Arak manager)
Dawood Agha Jani (Natanz official)	Ehsan Monajemi (Natanz construction manager)	Jafar Mohammadi (adviser to AEIO)
Ali Hajinia Leilabadi (director of Mesbah Energy)	Lt. Gen. Mohammad Mehdi Nejad Nouri (Malak Ashtar University of Defence Technology rector)	Gen Hosein Salimi (Commander, IRGC Air Force)
Ahmad Vahid Dastjerdi (head of Aerospace Industries Org., AIO)	Reza Gholi Esmaeli (AIO official)	Bahmanyar Morteza Bahmanyar (AIO official)
Maj. Gen. Yahya Rahim Safavi (Commander in Chief, IRGC)		
Entities Added by Resolution 1747		
Ammunition and Metallurgy Industries Group (controls 7th of Tir)	Esfahan Nuclear Fuel Research and Production Center and Esfahan Nuclear Technology Center	Kavoshyar Company (subsidiary of AEIO)

Parchin Chemical Industries (branch of DIO)	Karaj Nuclear Research Center	Novin Energy Company
Cruise Missile Industry Group	Bank Sepah (funds AIO and subordinate entities)	Sanam Industrial Group (subordinate to AIO)
Ya Mahdi Industries Group	Qods Aeronautics Industries (produces UAV's, para-gliders for IRGC asymmetric warfare)	Pars Aviation Services Company (maintains IRGC Air Force equipment)
Sho'a Aviation (produces IRGC light aircraft for asymmetric warfare)	Fereidoun Abbasi-Davani (senior defense scientist)	Mohasen Fakrizadeh-Mahabai (defense scientist)
Seyed Jaber Safdari (Natanz manager)	Amir Rahimi (head of Esfahan nuclear facilities)	Mohsen Hojati (head of Fajr Industrial Group)
Mehrdada Akhlaghi Ketabachi (head of SBIG)	Naser Maleki (head of SHIG)	Ahmad Derakshandeh (head of Bank Sepah)
Brig. Gen. Morteza Reza'i (Deputy commander-in-chief, IRGC)	Vice Admiral Ali Akbar Ahmadiyan (chief of IRGC Joint Staff)	Brig. Gen. Mohammad Reza Zahedi (IRGC ground forces commander)
Rear Admiral Morteza Safari (commander, IRGC Navy)	Brig. Gen. Mohammad Hejazi (Basij commander)	Brig. Gen. Qasem Soleimani (Qods Force commander)
Gen. Mohammad Baqr Zolqadr (IRGC officer serving as deputy Interior Minister		
Entities Added by Resolution 1803		
Thirteen Iranians named in Annex 1 to Resolution 1803; all reputedly involved in various aspects of nuclear program	Abzar Boresh Kaveh Co. (centrifuge production)	Barzaganin Tejaral Tavanmad Saccal
Electro Sanam Co.	Ettehad Technical Group (AIO front co.)	Industrial Factories of Precision
Jabber Ibn Hayan	Joza Industrial Co.	Khorasan Metallurgy Industries
Niru Battery Manufacturing Co. (Makes batteries for Iranian military and missile systems)	Pshgam (Pioneer) Energy Industries	Safety Equipment Procurement (AIO front, involved in missiles)
Tamas Co. (involved in uranium enrichment)		

Entities Designated Under U.S. Executive Order 13382 (many designations coincident with designations under U.N. resolutions)	
Entity	**Date Named**
Shahid Hemmat Industrial Group (Iran)	June 2005, Sept. 07
Shahid Bakeri Industrial Group (Iran)	June 2005, Feb. 2009
Atomic Energy Organization of Iran	June 2005
Novin Energy Company (Iran)	January 2006
Mesbah Energy Company (Iran)	January 2006

Table 10 (Continued)	
Four Chinese entities: Beijing Alite Technologies, LIMMT Economic and Trading Company, China Great Wall Industry Corp, and China National Precision Machinery Import/Export Corp.	June 2006
Sanam Industrial Group (Iran)	July 2006
Ya Mahdi Industries Group (Iran)	July 2006
Bank Sepah (Iran)	January 2007
Defense Industries Organization (Iran)	March 2007
Pars Trash (Iran, nuclear program)	June 2007
Farayand Technique (Iran, nuclear program)	June 2007
Fajr Industries Group (Iran, missile program)	June 2007
Mizan Machine Manufacturing Group (Iran, missile prog.)	June 2007
Aerospace Industries Organization (AIO) (Iran)	Sept. 2007
Korea Mining and Development Corp. (N. Korea)	Sept. 2007
Islamic Revolutionary Guard Corps (IRGC)	October 21, 2007
Ministry of Defense and Armed Forces Logistics	October 21, 2007
Bank Melli (Iran's largest bank, widely used by the Guard); Bank Melli Iran Zao (Moscow); Melli Bank PC (U.K.)	October 21, 2007
Bank Kargoshaee	October 21, 2007
Arian Bank (joint venture between Melli and Bank Saderat). Based in Afghanistan	October 21, 2007
Bank Mellat (provides banking services to Iran's nuclear sector); Mellat Bank SB CJSC (Armenia). Reportedly has $1.4 billion in assets in UAE	October 21, 2007
Persia International Bank PLC (U.K.)	October 21, 2007
Khatam ol Anbiya Gharargah Sazendegi Nooh (Revolutionary Guard construction, contracting arm, with $7 billion in oil, gas deals	October 21, 2007
Oriental Oil Kish (Iranian oil exploration firm)	October 21, 2007
Ghorb Karbala; Ghorb Nooh (synonymous with Khatam ol Anbiya)	October 21, 2007
Sepasad Engineering Company (Guard construction affiliate)	October 21, 2007
Omran Sahel (Guard construction affiliate)	October 21, 2007
Sahel Consultant Engineering (Guard construction affiliate)	October 21, 2007
Hara Company	October 21, 2007
Gharargahe Sazandegi Ghaem	October 21, 2007
Bahmanyar Morteza Bahmanyar (AIO, Iran missile official, see above under Resolution 1737)	October 21, 2007
Ahmad Vahid Dastjerdi (AIO head, Iran missile program)	October 21, 2007
Reza Gholi Esmaeli (AIO, see under Resolution 1737)	October 21, 2007
Morteza Reza'I (deputy commander, IRGC) See also Resolution 1747	October 21, 2007
Mohammad Hejazi (Basij commander). Also, Resolution 1747	October 21, 2007
Ali Akbar Ahmadian (Chief of IRGC Joint Staff). Resolution 1747	October 21, 2007
Hosein Salimi (IRGC Air Force commander). Resolution 1737	October 21, 2007
Qasem Soleimani (Qods Force commander). Resolution 1747	October 21, 2007
Future Bank (Bahrain-based but allegedly controlled by Bank Melli)	March 12, 2008
Yahya Rahim Safavi (former IRGC Commander in Chief	July 8, 2008
Mohsen Fakrizadeh-Mahabadi (senior Defense Ministry scientist)	July 8, 2008
Dawood Agha-Jani (head of Natanz enrichment site)	July 8, 2008
Mohsen Hojati (head of Fajr Industries, involved in missile program)	July 8, 2008
Mehrdada Akhlaghi Ketabachi (heads Shahid Bakeri Industrial Group)	July 8, 2008

Naser Maliki (heads Shahid Hemmat Industrial Group)	July 8, 2008
Tamas Company (involved in uranium enrichment)	July 8, 2008
Shahid Sattari Industries (makes equipment for Shahid Bakeri)	July 8, 2008
7th of Tir (involved in developing centrifuge technology)	July 8, 2008
Ammunition and Metallurgy Industries Group (partner of 7th of Tir)	July 8, 2008
Parchin Chemical Industries (deals in chemicals used in ballistic missile programs)	July 8, 2008
Karaj Nuclear Research Center	August 12, 2008
Esfahan Nuclear Fuel Research and Production Center (NFRPC)	August 12, 2008
Jabber Ibn Hayyan (reports to Atomic Energy Org. of Iran, AEIO)	August 12, 2008
Safety Equipment Procurement Company	August 12, 2008
Joza Industrial Company (front company for Shahid Hemmat Industrial Group, SHIG)	August 12, 2008
Islamic Republic of Iran Shipping Lines (IRISL) and 18 affiliates, including Val Fajr 8; Kazar; Irinvestship; Shipping Computer Services; Iran o Misr Shipping; Iran o Hind; IRISL Marine Services; Iriatal Shipping; South Shipping; IRISL Multimodal; Oasis; IRISL Europe; IRISL Benelux; IRISL China; Asia Marine Network; CISCO Shipping; and IRISL Malta	September 10, 2008
Firms affiliated to the Ministry of Defense, including Armament Industries Group; Farasakht Industries; Iran Aircraft Manufacturing Industrial Co.; Iran Communications Industries; Iran Electronics Industries; and Shiraz Electronics Industries	September 17, 2008
Export Development Bank of Iran. Provides financial services to Iran's Ministry of Defense and Armed Forces Logistics	October 22, 2008
Assa Corporation (alleged front for Bank Melli involved in managing property in New York City on behalf of Iran)	December 17, 2008
11 Entities Tied to Bank Melli: Bank Melli Iran Investment (BMIIC); Bank Melli Printing and Publishing; Melli Investment Holding; Mehr Cayman Ltd.; Cement Investment and Development; Mazandaran Cement Co.; Shomal Cement; Mazandaran Textile; Melli Agrochemical; First Persian Equity Fund; BMIIC Intel. General Trading	March 3, 2009
Entities Sanctioned Under Executive Order 13224 (Terrorism Entities)	
Qods Force	October 21, 2007
Bank Saderat (allegedly used to funnel Iranian money to Hezbollah, Hamas, PIJ, and other Iranian supported terrorist groups)	October 21, 2007
Al Qaeda Operatives in Iran: Saad bin Laden; Mustafa Hamid; Muhammad Rab'a al-Bahtiyti; Alis Saleh Husain	Jan. 16, 2009
Entities Sanctioned Under the Iran Non-Proliferation Act and other U.S. Proliferation Laws	
Norinco (China). For alleged missile technology sale to Iran.	May 2003
Taiwan Foreign Trade General Corporation (Taiwan)	July 4, 2003
Tula Instrument Design Bureau (Russia). For alleged sales of laser-guided artillery shells to Iran.	September 17, 2003
13 entities sanctioned including companies from Russia, China, Belarus, Macedonia, North Korea, UAE, and Taiwan.	April 7, 2004
14 entities from China, North Korea, Belarus, India (two nuclear scientists, Dr. Surendar and Dr. Y.S.R. Prasad), Russia, Spain, and Ukraine.	September 29, 2004
14 entities, mostly from China, for alleged supplying of Iran's missile program. Many, such as	December 2004

Table 10 (Continued)	
North Korea's Changgwang Sinyong and China's Norinco and Great Wall Industry Corp, have been sanctioned several times previously. Newly sanctioned entities included North Korea's Paeksan Associated Corporation, and Taiwan's Ecoma Enterprise Co.	and January 2005
9 entities, including those from China (Norinco yet again), India (two chemical companies), and Austria. Sanctions against Dr. Surendar of India (see September 29, 2004) were ended, presumably because of information exonerating him.	December 26, 2005
7 entities. Two Indian chemical companies (Balaji Amines andPrachi Poly Products); two Russian firms (Rosobornexport and aircraft manufacturer Sukhoi); two North Korean entities (Korean Mining and Industrial Development, and Korea Pugang Trading); and one Cuban entity (Center for Genetic Engineering and Biotechnology).	August 4, 2006
9 entities. Rosoborneskport, Tula Design, and Komna Design Office of Machine Building, and Alexei Safonov (Russia); Zibo Chemical, China National Aerotechnology, and China National Electrical (China). Korean Mining and Industrial Development (North Korea) for WMD or advanced weapons sales to Iran (and Syria).	January 2007
14 entities, including Lebanese Hezbollah. Some were penalized for transactions with Syria. Among the new entities sanctioned for assisting Iran were Shanghai Non-Ferrous Metals Pudong Develop-ment Trade Company (China); Iran's Defense Industries Organiza-tion; Sokkia Company (Singapore); Challenger Corporation (Malaysia); Target Airfreight (Malaysia); Aerospace Logistics Services (Mexico); and Arif Durrani (Pakistani national).	April 23, 2007
13 entities: China Xinshidai Co.; China Shipbuilding and Offshore International Corp.; Huazhong CNC (China); IRGC; Korea Mining Development Corp. (North Korea); Korea Taesong Trading Co. (NK); Yolin/Yullin Tech, Inc. (South Korea); Rosoboronexport (Russia sate arms export agency); Sudan Master Technology; Sudan Technical Center Co; Army Supply Bureau (Syria); R and M International FZCO (UAE); Venezuelan Military Industries Co. (CAVIM);	October 23, 2008
Entities Designated as Threats to Iraqi Stability under Executive Order 13438	
Ahmad Forouzandeh. Commander of the Qods Force Ramazan Headquarters, accused of fomenting sectarian violence in Iraq and of organizing training in Iran for Iraqi Shiite militia fighters	January 9, 2008
Abu Mustafa al-Sheibani. Iran based leader of network that funnels Iranian arms to Shiite militias in Iraq.	January 9, 2008
Isma'il al-Lami (Abu Dura). Shiite militia leader, breakaway from Sadr Mahdi Army, alleged to have committed mass kidnapings and planned assassination attempts against Iraqi Sunni politicians	January 9, 2008
Mishan al-Jabburi. Financier of Sunni insurgents, owner of pro-insurgent Al-Zawra television, now banned	January 9, 2008
Al Zawra Television Station	January 9, 2008

Status of Some U.S.-Iran Assets Disputes

Iranian leaders continue to assert that the United States is holding Iranian assets, and that this is an impediment to improved relations. A U.S.-Iran Claims Tribunal at the Hague continues to arbitrate cases resulting from the 1980 break in relations and freezing of some of Iran's assets.

Major cases yet to be decided center on hundreds of Foreign Military Sales (FMS) cases between the United States and the Shah's regime, which Iran claims it paid for but were unfulfilled. About $400 million in proceeds from the resale of that equipment was placed in a DOD FMS account, and about $22 million in Iranian diplomatic property remains blocked, although U.S. funds have been disbursed—credited against the DOD FMS account—to pay judgments against Iran for past acts of terrorism against Americans. Other disputes include the mistaken U.S. shoot-down on July 3, 1988, of an Iranian Airbus passenger jet (Iran Air flight 655), for which the United States, in accordance with an ICJ judgment, paid Iran $61.8 million in compensation ($300,000 per wage earning victim, $150,000 per non-wage earner) for the 248 Iranians killed. The United States has not compensated Iran for the airplane itself. As it has in past similar cases, the Administration has opposed a terrorism lawsuit against Iran by victims of the U.S. Embassy Tehran seizure on the grounds of diplomatic obligation.[68]

CONCLUSION

Mistrust between the United States and Iran's Islamic regime has run deep almost three decades and many argue that it is unlikely to be quickly overcome, even if the Obama Administration initiates—and Iran accepts—comprehensive direct talks with Iran. As noted, that possibility might have been made more remote by the violent dispute over the June 12 presidential election in Iran. Despite the internal power struggle, many experts say that all factions in Iran are united on major national security issues and that U.S.-Iran relations might not improve unless or until the Islamic regime is removed or moderates substantially, even if a nuclear deal is reached and implemented. Others say that, despite Ahmadinejad's presidency, the United States and Iran have a common long term interest in stability in the Persian Gulf and South Asia regions in the aftermath of the defeat of the Taliban and the regime of Saddam Hussein and that major diplomatic overtures might now yield fruit.

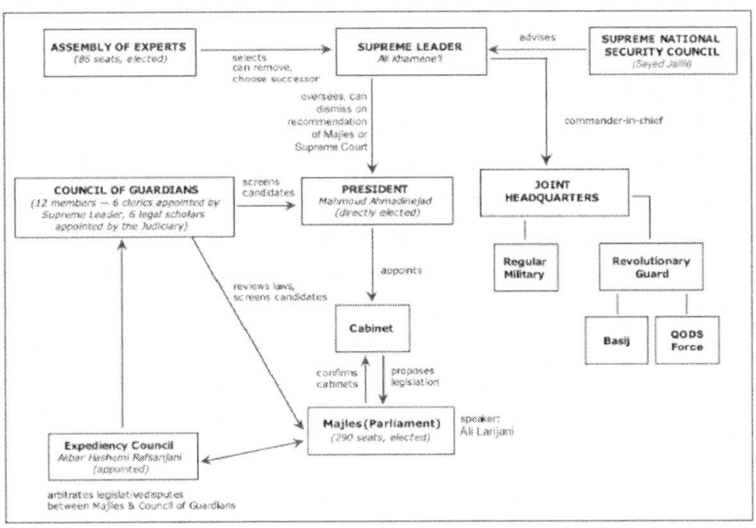

Figure 1. Structure of the Iranian Government

Source: Map Resources. Adapted by CRS (April 2005).

Figure 2. Map of Iran

End Notes

[1]The Assembly also has the power to amend Iran's constitution.

[2] The Council of Guardians consists of six Islamic jurists and six secular lawyers. The six Islamic jurists are appointed by the Supreme Leader. The six lawyers on the Council are selected by the judiciary but confirmed by the Majles.

[3] Rafsanjani was constitutionally permitted to run because a third term would not have been consecutive with his previous two terms. In the 2001 presidential election, the Council permitted 10 out of the 814 registered candidates.

[4] A paper published by Chatham House and the University of St. Andrews strongly questions how Ahmadinejad's vote could have been as large as reported by official results, in light of past voting patterns throughout Iran. "Preliminary Analysis of the Voting Figures in Iran's 2009 Presidential Election." www.chathamhouse.org.uk.

[5] Other names by which this group is known is the Mojahedin-e-Khalq Organization (MEK or MKO) and the National Council of Resistance (NCR).

[6] The designation was made under the authority of the Anti-Terrorism and Effective Death Penalty Act of 1996 (P.L. 104-132).

[7] Kampeas, Ron. "Iran's Crown Prince Plots Nonviolent Insurrection from Suburban Washington." Associated Press, August 26, 2002.

[8]See http://www.whitehouse.gov/nsc/nss/2006/.

[9] For a more extensive discussion of the IRGC, see Katzman, Kenneth. "The Warriors of Islam: Iran's Revolutionary Guard," *Westview Press*, 1993.

[10]Text at http://www.dni.gov/press_releases/20071203_release.pdf.

[11] In November 2006, the IAEA, at U.S. urging, declined to provide technical assistance to the Arak facility on the grounds that it was likely for proliferation purposes.

[12] Lancaster, John and Kamran Khan. "Pakistanis Say Nuclear Scientists Aided Iran." *Washington Post*, January 24, 2004.

[13] For Iran's arguments about its program, see Iranian paid advertisement "An Unnecessary Crisis—Setting the Record Straight About Iran's Nuclear Program," in the *New York Times*, November 18, 2005. P. A11.

[14]Stern, Roger. "The Iranian Petroleum Crisis and United States National Security," *Proceedings of the National Academy of Sciences of the United States of America*. December 26, 2006.

[15] For text of the agreement, see http://www.iaea.org/NewsCenter/Focus/IaeaIran/eu_iran14112004.shtml. EU-3—Iran negotiations on a permanent nuclear pact began on December 13, 2004, and related talks on a trade and cooperation accord (TCA) began in January 2005.

[16] In November 2006, the IAEA, at U.S. urging, declined to provide technical assistance to the Arak facility on the grounds that it was likely for proliferation purposes.

[17]Voting in favor: United States, Australia, Britain, France, Germany, Canada, Argentina, Belgium, Ghana, Ecuador, Hungary, Italy, Netherlands, Poland, Portugal, Sweden, Slovakia, Japan, Peru, Singapore, South Korea, India. Against: Venezuela. Abstaining: Pakistan, Algeria, Yemen, Brazil, China, Mexico, Nigeria, Russia, South Africa, Sri Lanka, Tunisia, and Vietnam.

[18]Voting no: Cuba, Syria, Venezuela. Abstaining: Algeria, Belarus, Indonesia, Libya, South Africa.

[19]See http://daccessdds.un.org/doc/UNDOC/GEN/N06/290/88/PDF/N0629088.pdf?OpenElement.

[20] One source purports to have obtained the contents of the package from ABC News: http://www.basicint.org/pubs/Notes/BN060609.htm.

[21]Dempsey, Judy. "U.S. Urged to Talk With Iran." *International Herald Tribune*, February 5, 2009.

[22] Annual Threat Assessment of the Intelligence for the Senate Select Committee on Intelligence, Dennis C. Blair, Director of National Intelligence, February 12, 2009.

[23] Broad, William and David Sanger. "Relying On Computer, U.S. Seeks to Prove Iran's Nuclear Aims." *New York Times*, November 13, 2005.

[24] U.S. Department of State. *Country Reports on Terrorism 2007*. Released April 30, 2008. http://www.state.gov/s/ct/ rls/crt/2007/103711.htm.

[25] Walsh, Elsa. "Annals of Politics: Louis Freeh's Last Case." *The New Yorker*, May 14, 2001. The June 21, 2001, federal grand jury indictments of 14 suspects (13 Saudis and a Lebanese citizen) in the Khobar bombing indicate that Iranian agents may have been involved, but no indictments of any Iranians were announced. In June 2002, Saudi Arabia reportedly sentenced some of the eleven Saudi suspects held there. The 9/11 Commission final report asserts that Al Qaeda might have had some as yet undetermined involvement in the Khobar Towers attacks.

[26] This issue is covered in greater depth in CRS Report RS22323, *Iran's Activities and Influence in Iraq*, by Kenneth Katzman.

[27]CNN "Late Edition" interview with Hamas co-founder Mahmoud Zahar, January 29, 2006.

[28]Hamas Leader Praises Iran's Help in Gaza 'Victory.' CNN.com. February 1, 2009.

[29] Slackman, Michael. "Egypt Accuses Hezbollah of Plotting Attacks in Sinai and Arms Smuggling to Gaza." New York Times, April 14, 2009

[30] Hezbollah is believed responsible for the October 1983 bombing of the U.S. Marine barracks in Beirut, as well as attacks on U.S. Embassy Beirut facilities in April 1983 and September 1984, and for the hijacking of TWA Flight 847 in June 1985 in which Navy diver Robert Stetham was killed. Hezbollah is also believed to have committed the March 17, 1992, bombing of Israel's embassy in that city, which killed 29 people. Its last known terrorist attack outside Lebanon was the July 18, 1994, bombing of a Jewish community center in Buenos Aires, which killed 85. On October 31, 2006, Argentine prosecutors asked a federal judge to seek the arrest of Rafsanjani, former Intelligence Minister Ali Fallahian, former Foreign Minister Ali Akbar Velayati, and four other Iranian officials for this attack.

[31] "Israel's Peres Says Iran Arming Hizbollah." Reuters, February 4, 2002.

[32]Rotella, Sebastian. "In Lebanon, Hezbollah Arms Stockpile Bigger, Deadlier." *Los Angeles Times*, May 4, 2008.

[33]Shadid, Anthony. "Armed With Iran's Millions, Fighters Turn to Rebuilding." *Washington Post*, August 16, 2006.

[34]See CRS Report RL30588, *Afghanistan: Post-Taliban Governance, Security, and U.S. Policy*, by Kenneth Katzman.

[35]Gertz, Bill. "Al Qaeda Terrorists Being Held by Iran." *Washington Times*, July 24, 2003.

[36]Keto, Alex. "White House Reiterates Iran Is Harboring Al Qaeda." *Dow Jones Newswires*, May 19, 2003.

[37] Gertz, Bill. "CIA Points to Continuing Iran Tie to Al Qaeda." *Washington Times*, July 23, 2004.

[38] "Tehran Pledges to Crack Down on Militants." Associated Press, July 18, 2005.

[39]Arostegui, Martin. "Uruguay Caught Buying Iran Arms." *Washington Times*, October 12, 2007.

[40] An exception was the abortive 1985-1986 clandestine arms supply relationship with Iran in exchange for some American hostages held by Hezbollah in Lebanon (the so-called "Iran-Contra Affair"). Iran has an interest section in Washington D.C. under the auspices of the Embassy of Pakistan; it is staffed by Iranian-Americans. The U.S. interest section in Tehran has no American personnel; it is under the Embassy of Switzerland.

[41] Sciolino, Elaine. *The Outlaw State: Saddam Hussein's Quest for Power and the Gulf Crisis*. New York: John Wiley and Sons, 1991. p. 168.

[42] Cooper, Helene and David Sanger. "Strategy on Iran Stirs New Debate at White House." *New York Times*, June 16, 2007.

[43] The FY2007 defense authorization law (P.L. 109-364) called for a report by the Administration on all aspects of U.S. policy and objectives on Iran (and required the DNI to prepare a national intelligence estimate on Iran, which was released on December 3, 2007 as discussed above).

[44]Wright, Robin. "U.S. In 'Useful' Talks With Iran." *Los Angeles Times*, May 13, 2003.

[45] For an extended discussion of U.S. air strike options on Iran, see Rogers, Paul. *Iran: Consequences Of a War*. Oxford Research Group, February 2006.

[46] See also, Washington Institute for Near East Policy. The Last Resort: Consequences of Preventive Military Action Against Iran, by Patrick Clawson and Michael Eisenstadt. June 2008.

[47] Shanker, Thom. "U.S. and Britain to Add Ships to Persian Gulf in Signal to Iran," *New York Times*, December 21, 2006.

[48] "New Persian Gulf Security Effort Expected to Fuel Arms Sales in FY-07." *Inside the Pentagon*, November 9, 2006.

[49] CRS conversations with U.S. officials responsible for Iran policy. 1980-1990. After a period of suspension of such assistance, in 1995, the Clinton Administration accepted a House-Senate conference agreement to include $18-$20 million in funding authority for covert operations against Iran in the FY1996 Intelligence Authorization Act (H.R. 1655, P.L. 104-93), according to a *Washington Post* report of December 22, 1995. The Clinton Administration reportedly focused the covert aid on changing the regime's behavior, rather than its overthrow.

[50] Ross, Brian and Richard Esposito. Bush Authorizes New Covert Action Against Iran. http://blogs.abcnews.com/theblotter/2007/05/bush_authorizes.html.

[51] This legislation was a modification of H.R. 282, which passed the House on April 26, 2006, by a vote of 397-21, and S. 333, which was introduced in the Senate.

[52] Three other Iranian Americans were arrested and accused by the Intelligence Ministry of actions contrary to national security in May 2007: U.S. funded broadcast (Radio Farda) journalist Parnaz Azima (who was not in jail but was not allowed to leave Iran); Kian Tajbacksh of the Open Society Institute funded by George Soros; and businessman and peace activist Ali Shakeri. Several congressional resolutions called on Iran to release Esfandiari (S.Res. 214 agreed to by the Senate on May 24; H.Res. 430, passed by the House on June 5; and S.Res. 199). All were released by October 2007.

[53]Stockman, Farah. "'Long Struggle' With Iran Seen Ahead." *Boston Globe*, March 9, 2006.

[54] Berlin Says U.S. and France Guilty of Hypocrisy. Spiegel Online, September 24, 2007. http://www.spiegel.de/international/world/0,1518,507443,00.html

[55] During the active period of talks, which began in December 2002, there were working groups focused not only on the TCA terms and proliferation issues but also on Iran's human rights record, Iran's efforts to derail the Middle East peace process, Iranian-sponsored terrorism, counter-narcotics, refugees, migration issues, and the Iranian opposition PMOI.

[56] On November 14, 1979, President Carter declared a national emergency with respect to Iran, renewed every year since 1979.

[57] "The Fight Over Letting Foreigners Into Iran's Oilfields." *The Economist*, July 14, 2001.

[58]Such laws include the Atomic Energy Act of 1954 and the Energy Policy Act of 2005 (P.L. 109-58).

[59] The provision contains certain exceptions to ensure the safety of astronauts, but it nonetheless threatened to limit U.S. access to the international space station after April 2006, when Russia started charging the United States for transportation on its Soyuz spacecraft. Legislation in the 109th Congress (S. 1713, P.L. 109-112) amended the provision in order to facilitate continued U.S. access and extended INA sanctions provisions to Syria.

[60] Warrick, Joby. "Iran Using Fronts to Get Bomb Parts From U.S." *Washington Post*, January 11, 2009; Institute for Science and International Security. "Iranian Entities' Illicit Military Procurement Networks." David Albright, Paul Brannan, and Andrea Scheel. January 12, 2009.

[61]Mufson, Steven and Robin Wright. "Iran Adapts to Economic Pressure." *Washington Post,* October 29, 2007.

[62]Kessler, Glenn. "U.S. Moves to Isolate Iranian Banks." *Washington Post*, September 9, 2006.

[63] An August 1997 amendment to the trade ban (Executive Order 13059) prevented U.S. companies from knowingly exporting goods to a third country for incorporation into products destined for Iran.

[64] "Iran Says Halliburton Won Drilling Contract." *Washington Times*, January 11, 2005.

[65] Prada, Paulo, and Betsy McKay. Trading Outcry Intensifies. *Wall Street Journal*, March 27, 2007; Brush, Michael. Are You Investing in Terrorism? MSN Money, July 9, 2007.

[66] Originally called the Iran-Libya Sanctions Act, or ILSA; P.L. 104-172, August 5, 1996. It was renewed by P.L. 107-24, August 3, 2001; renewed again for two months by P.L. 109-267; and renewed and amended by P.L. 109-293.

[67] For information on the steps taken by individual states, see National Conference of State Legislatures. State Divestment Legislation.

[68]See CRS Report RL31258, *Suits Against Terrorist States by Victims of Terrorism*, by Jennifer K. Elsea.

In: Iran Social, Economic and Political Developments
Editor: Joshua A. Owenstein

ISBN: 978-1-60741-552-7

Chapter 3

IRAN'S 2009 PRESIDENTIAL ELECTIONS*

Casey L. Addis

SUMMARY

On June 12, 2009, following a heated campaign between reformist candidate Mir Hussein Musavi and incumbent President Mahmoud Ahmadinejad, Iranians turned out in record numbers to vote in the presidential election. Shortly after the polls closed, the Interior Minister announced that President Ahmadinejad had been reelected by a 62% margin. The announcement was followed by allegations of vote rigging and election fraud and prompted supporters of leading reformist candidate Mir Hussein Musavi and others to hold public demonstrations in several major cities of a size and intensity unprecedented since the Iranian Revolution of 1979.

Despite a government ban on unauthorized public gatherings, protests have reportedly continued every day since the election. Restrictions on foreign and domestic journalists, reported disruptions of mobile phone networks, limited accessibility of some internet sites, mass arrests, and clashes between civilian protestors and Basij forces have garnered international attention and increased concerns about the Iranian government's apparent disregard for human rights and basic civil liberties.

Regardless of the actual election results, the current stand-off between the government and opponents of the election outcome has caused observers to speculate about how this stalemate will be resolved, and what the outcome might mean for U.S. efforts to resolve the issue of Iran's nuclear weapons program, its support for terrorism, and other national security concerns.

* This is an edited, reformatted and augmented version of a CRS Report for Congress publication dated June 2009.

Recent Developments

In his speech on Friday, June 19, Supreme Leader Khamenei demanded an end to the protests, reiterated his support for President Ahmadinejad, and accused foreign "enemies" of interfering in Iran's domestic affairs. Protests continued in Tehran and in other cities, however, and on Saturday, June 20, Iranian Basij and Revolutionary Guard forces reportedly used tear gas and live ammunition to disperse crowds. Ten deaths were reported, bringing the unofficial toll to at least 17, although many speculate that violence between police and military forces and the protestors may be more widespread and lethal than media reports indicate. On June 22, reports indicated that the Basij and Revolutionary Guard have been deployed throughout Tehran as the government crackdown on demonstrations continues to intensify. In addition, the Iranian government also appears to be continuing its arrests of reformist leaders. On June 21, members of former President Rafsanjani's family were reportedly arrested, causing speculation that rifts in Iran's religious leadership could be widening.[1]

On Saturday, June 20, the Guardian Council held a meeting with all presidential candidates to discuss the election outcome and fraud allegations. Reformist candidate Mir Hussein Musavi did not attend the meeting on Saturday on the grounds that he has already rejected any solution to the current stand-off other than a new election.[2] Reports circulated on Monday, June 22 that the number of ballots cast in at least 50 voting stations as reported by the government exceeds the number of registered voters in that area. The Guardian Council acknowledged the "irregularities," but insisted that they in no way would have changed the election outcome.[3]

As information leaks out of Iran despite a government ban on foreign journalists and restricted access to some internet sites, it is becoming clear to some observers that the Iranian government may have chosen to impose the election outcome by force. Images and available information have heightened international concerns that the human rights situation in Iran is further deteriorating, and have raised questions about the extent of the Iranian government's campaign to end the protests and silence the opposition.

In response to increased violence against the Iranian protesters and amid some criticisms of his prior remarks, President Obama called upon the government of Iran to "stop all violent and unjust actions against its own people" and stated that, "The Iranian people will ultimately judge the actions of their own government. If the Iranian government seeks the respect of the international community, it must respect the dignity of its own people and govern through consent, not coercion."[4] Many observers believe that President Obama is attempting to balance the need to condemn the violence against the protestors with the need to avoid the perception of U.S. interference, which some worry could prompt the Iranian government to clamp down further on freedom of expression as well as jeopardize U.S. efforts to engage Iran on the issue of its nuclear program.

Iran's 2009 Presidential Election

The reported outcome of the June 12, 2009 presidential election in Iran prompted public demonstrations in several major cities of a size and intensity unprecedented since the Iranian

Revolution of 1979. The announcement that President Mahmoud Ahmadinejad was reelected by a 62% margin was followed by allegations of vote rigging and election fraud. Supporters of leading reformist candidate Mir Hussein Musavi and others continue to stage large protests in the streets of Tehran and other major cities that have drawn international attention. The actions taken by the Iranian government in the hours following the election, ongoing demonstrations, arrests, and clashes between civilians and Iran's paramilitary Basij forces have led some observers to argue that Iran's political dynamics have shifted considerably, and that the relationship between the government of the Islamic Republic and its citizens has been fundamentally shaken. While some analysts have speculated about military coups, color revolutions, and the future of the Iranian regime, others have reserved judgment about the likelihood of various potential short term developments because of the unpredictability inherent in the dynamic situation prevailing across Iran at this time. Iran's institutions and centralized decision making are opaque, and outside observation and analysis is further hampered by the fact that the Iranian government has restricted access to foreign and domestic journalists and interrupted the availability of mobile phone service and various internet sites and services, limiting access to reliable information on the situation in Iran and compounding international concerns.

The short term effects of the election and its aftermath on Iran's political system and social contract are difficult to foresee. It is not clear at this point whether widespread public anger over the election outcome will lead to a wider challenge of the core principles of the Islamic Republic, and much may depend on the decisions of officials, candidates, and citizens in the days and weeks ahead. Regardless of the short term outcome, many analysts agree that the events surrounding the 2009 Iranian presidential elections have upset the balance between the official and civil spheres of Iranian society and will have long term implications for both the government and the people of Iran and for U.S. policy.[5]

CANDIDATES AND CAMPAIGNS

In 2009, nearly 500 candidates for Iran's presidency filed their candidacy with the Guardian Council.[6] On May 20, 2009, the council announced that four candidates had been approved: incumbent president Mahmoud Ahmadinejad, conservative Mohsen Reza'i, reformist Mir Hussein Musavi, and reformist Mehdi Karrubi.

Social and political restrictions are often eased in Tehran during campaign season, but observers remarked that public activity this year was notably more energetic than would be expected. Some attribute this shift to the four years of crackdowns on social freedoms that have characterized President Ahmadinejad's term in office.[7] Others attribute the pre-election atmosphere to increased public tension between the candidates in the days leading up to the election, when the campaign became increasingly acrimonious.

During the week of June 3, 2009, the candidates participated in six live debates. The debate between incumbent President Mahmoud Ahmadinejad and reformist candidate Mir Hussein Musavi was particularly heated, most notably because of Ahmadinejad's open criticism of Musavi's wife, Zahra Rahnavard.[8] The debates offered the public an opportunity to observe fierce exchanges between the candidates for the first time in a presidential election and reportedly were watched by 40 to 50 million viewers, according to Iranian media reports.[9]

On June 9, Ali Akbar Hashemi Rafsanjani, a former president who now heads two powerful oversight bodies, issued an open letter complaining about the silence of Khamenei following the "insults, lies, and false allegations" by Ahmadinejad during the campaign debates.[10] It is rare in Iran for senior leaders to publicly criticize the supreme leader and many observers viewed the letter as a reflection of the intensity of the campaigns. Others perceive that Rafsanjani, who is often at odds with the Khamenei and lost the 2005 presidential race to Ahmdinejad, may be interested in forming an alternate power center in the government.

Musavi appeared to experience a surge in public support in the final days of the campaign. The night before the election, on June 11, Musavi supporters reportedly formed a human chain on a main thoroughfare through Tehran.[11] The mass rally prompted some analysts and observers to speculate that Ahmadinejad's chances at reelection were dwindling. Some reports that the rural and urban poor population of Iran was shifting its support away from Ahmadinejad also surfaced in the week before the election. Many attributed this to increases in inflation and unemployment, compounded by international sanctions and the global financial crises that had disproportionately affected the poor, despite increases in wages and pensions provided under the Ahmadinejad government.[12]

The large rallies in favor of Musavi during the last days of the campaign may have caused alarm among some factions of the Iranian government.[13] Prior to the election, Iranian Revolutionary Guard Corps (IRGC) commander General Mohammad Ja'fari publicly stated that any attempt at a velvet revolution in Iran would be crushed.[14] The statement further fueled speculation that the regime felt threatened by the apparent popularity of Musavi in the last days of the campaign.

As observers watched the campaign unfold, most predicted a close race between Musavi and Ahmadinejad and many anticipated that a run-off would be necessary to determine a winner. Many observers also agreed that voter turnout may tip the election in favor of Musavi. During past elections, low voter turnout has been due in part to boycotts on the part of reform-minded Iranians, including many Iranian Americans who are eligible to vote. The Iranian system, in which the Guardian Council chooses which candidates are eligible to run, has in the past led some Iranians to feel that they have no attractive choice among the candidates.

Election and Results

On June 12, following the heated campaign between Musavi and Ahmadinejad, Iranians went to the polls. Record voter turnout was reported throughout the day and the Interior Ministry ordered that voting centers stay open to accommodate those waiting to vote.[15] Many observers were optimistic that pro-reform segments of the population, who had boycotted elections in the past, had gone to the polls in favor of Musavi. Large campaign rallies prior to the election had even sparked discussion of a possible "Green Revolution." As the polls closed, however, the prospects began to dim for a Musavi victory, and for a popularly-accepted election outcome. As the polls closed, police and paramilitary (Basij) forces reportedly were deployed throughout Tehran, locking down the Interior Ministry where votes were being counted. Internet sites and mobile phones were also reportedly disabled. Less than three hours after the polls closed, the Interior Ministry announced that the election results were in and that Mahmoud Ahmadinejad had won, capturing 62% of the vote. The Interior Ministry also reported that 39 million votes were cast (about 85% of Iran's eligible voters), an unprecedented turnout.[16]

Following the announcement by the Interior Ministry, Supreme Leader Ayatollah Ali Khamenei issued a statement congratulating President Ahmadinejad, which most observers interpreted as a certification of the election results. Khamenei said the "miraculous hand of God" was evident in the "great epic" of the election.[17] Both Ahmadinejad and Musavi claimed victory as the announcement was made, even before the Guardian Council certified the results.

Allegations of Fraud

Since no independent international observers were present for Iran's elections, it is difficult to ascertain the extent of alleged vote rigging or election violations that may have taken place. The expulsion of most foreign journalists from Iran and the government's interruption of mobile and internet communication have further complicated efforts to gain a clear picture of the events surrounding the election and its aftermath.

While many democracy promotion groups and NGOs have criticized the elections process in Iran, Iran's election procedures have been relatively well codified and in force for decades.[18] However, in the 2009 presidential election, doubts about the wide margin of victory reported for Ahmadinejad have overshadowed a number of serious procedural irregularities reported on election day, which are the subject of the formal complaints filed by the defeated candidates.

The Guardian Council has reportedly received and is investigating nearly 650 poll complaints from the three losing candidates. Musavi's formal complaints include: the heads of governors' offices sabotaged the issuance of identification cards to electoral observers before the ballots were collected and counted manually; ballot papers were not distributed properly in Shiraz and Tabriz; additional television campaigning was allowed for Ahmadinejad; the headlines of agencies including Rajanews, Fars, and IRNA focused on Ahmadinejad's victory in the election; and Article 40 of the Criminal Code regarding army-related crimes was broken through the involvement of Basij members in Ahmadinejad's campaign meetings.[19] Karrubi has expressed similar concerns.

Conservative candidate Mohsen Reza'i has also filed formal complaints, which some have perceived as an indication that it is not just reformist candidates and supporters who are dissatisfied with the results. According to the official result totals he received 678,000. He argues that he received between at least 5.3 and 7 million votes and that some estimates could be as high as 9 million. He says that his claim is based on his observers' reports from polling stations, information collected from 1,000 ballot boxes by his electoral headquarters, official opinion polls held in the country, and remarks addressed to him by voters and members of his electoral headquarters.[20]

Taken together, doubts about the margin of victory and concerns over perceived fraud stemming from procedural violations are fueling the largest protests since the Iranian Revolution and creating international concerns about the legitimacy of the election results. Initially, the candidates, the Iranian people, and many who followed the election expressed doubts that the record 39 million votes cast could have been counted in such a short time, especially in light of reports that voting times were extended. Others have pointed to additional irregularities. While individual reports are difficult to verify, some reports suggest that, in some cases, ballot boxes were picked up by Interior Ministry officials before polling places closed which would mark a change in election procedures whereby ballots are typically counted by officials at the local level.[21]

Khamenei urged the Guardian Council on June 15 to examine seriously the allegations of vote rigging, but urged the candidates to pursue their complaints through legal channels. He went on to state that the probe into vote-rigging allegations would be completed by June 25.[22] Musavi has said that invalidating the election is the only way to regain the people's trust in the regime and rejected outright the Guardian Council's offer to recount some of the votes.[23] Some observers dismissed the investigation into the election results as an attempt to provide a cooling off period for the demonstrators and dissatisfied public, rather than a legitimate review of the results. The Council has reportedly invited all the presidential candidates to a meeting scheduled to be held on Saturday, June 20; it is unclear whether one or all will attend.[24]

Some observers argue that the election results could be valid, despite the appearance of irregularities. They support this assertion with the claim that the young, liberal demographic in Iran is much smaller than it is often portrayed and that Ahmadinejad enjoys widespread support among the rural and urban poor, a more significant group in terms of size. These analysts also question the reliability of the polls prior to the election that indicated a close race. Others say that Ahmadinejad is often underestimated, as he was in 2005, and that his message of piety and anti-corruption coupled with his hard line on national security issues are both popular among the majority of Iranians. Others have argued that election fraud on such a massive scale would have involved many levels of the government and would be difficult to perpetrate and conceal.[25]

Other analysts assert that the allegations of fraud are likely true, and that the regime had motivation to interfere with the results. These analysts argue that the Iranian government might have felt sufficiently threatened by the success of Ahmadinejad's reformist opponents to mobilize a segment of the population that in large part boycotted the last elections. Some analysts have speculated that Khamenei engineered the election results in Ahmadinejad's favor so drastically in an effort to avoid a close election that could have been contested. Others argue that Khamenei wanted to send a political message to the U.S. and others that overtures to the Iranian public did not sway Iran from its ideology—a commitment to the nuclear program and an approval of Ahmadinejad's inflammatory rhetoric about Israel—and that discussions with the U.S. are not perceived by the Iranian people as a prize to be won. Some agree that Khamenei miscalculated, either by misjudging popular opinion or out of paranoia over suspected regime change efforts on the part of the U.S. and the West.[26]

Regardless of the actual election results, the public demonstrations on election night and continued protests in major cities across Iran have caused observers to speculate about how the current stand-off between the government and Musavi's supports will be resolved, and what the outcome might mean for U.S. efforts to resolve the issue of Iran's nuclear weapons program, its support for terrorism, and other national security concerns.

AFTERMATH

Demonstrations in Iran

Shortly after the election results were announced, Iran's interior ministry issued a ban on unauthorized public gatherings.[27] Despite the warning, protests have reportedly continued every day since the election in Tehran and other major cities—including Mashhad, Tabriz, Shiraz, and Isfahan.[28] Restrictions on journalists and government efforts to restrict telecommunications have made it difficult to know the scope of the public protests, but most accounts indicate numbers in the hundreds of thousands or more in Tehran. A counter demonstration in support of President Ahmadinejad also was reported, but most estimates indicate that it was significantly smaller than those in protest of the results—less than 10,000 people. Some media outlets alleged that the images of the Ahmadinejad rally were doctored to inflate the apparent size of the crowd.[29] Reports of arrests, injuries, and deaths are also difficult to substantiate, but have gained international attention and raised concerns about how the current stand-off between the government and the demonstrators might end. As the momentum reportedly continues, many observers are making comparisons with the days leading up to the Iranian revolution, and others express concerns that Iran might be nearing a "Tiananmen moment," in which the government ends protests with a massive show of force.[30]

At least 100 reformists reportedly were arrested on June 14 as public unrest mounted over the election results. Some of them reportedly have been released. On Sunday June 14, Iran's acting police chief, Ahmadreza Radan, gave the state press service an update on the arrests of protesters, and assured the public that "in the interrogation of related rebels, we intend to find the link between the plotters and foreign media."[31]

Musavi's first public appearance after the election was at a rally in Tehran's Revolution Square on Monday June 15, ending rumors of house arrest that circulated when he failed to show up for a planned press conference on June 13. The demonstrations were reportedly peaceful with no visible police presence. As the peaceful demonstration ended, a group was reported to attempt to set fire to the Basij compound away from the heart of the gathering. Basij members opened fire on the crowd from the roof of the facility. At least seven people were reportedly killed in the clash, and reports have since surfaced of other deaths and injuries, although the exact count is unknown.[32]

On the morning of June 15, fighting was reported at Tehran University, where students reportedly threw rocks and bricks at police and set fire to vehicles. The police reportedly used tear gas and plastic bullets to disperse the protesters, and reportedly raided university dormitories on both nights. Majles (Parliament) Speaker Ali Larijani condemned the raids, placing the blame on the Interior Ministry. Additional overnight raids were reported by the BBC on June 17.[33]

According to the BBC, on June 16 as night fell, residents of Tehran shouted protest messages from their rooftops, a scene "not witnessed since the final days of the Shah." On June 18, demonstrators staged a silent protest, mourning those who were injured and lost in the earlier protests.[34]

In addition to Musavi, other reformist leaders have joined the protests against the election outcome. Former President Mohammad Khatami's moderate clerical group, the Association of Combatant Clergy, issued a statement warning that "if this process becomes the norm, the republican aspect of the regime will be damaged and people will lose confidence in the system." Khatami also apologized to the Iranian people for his failure to protect their votes.[35]

On June 19, Supreme Leader Khamenei spoke at Friday prayers. In his address, he demanded an end to the protests and stated that political leaders would be held responsible for any violence. While he did not announce that the probe into the allegations of election fraud was completed, he did express his support for Ahmadinejad and seemed to reiterate his commitment to the election outcome stating that "there is 11 million votes difference...How can one rig 11 million votes?" He went on to criticize Great Britain and other "enemies" of Iran, stating that: "Some of our enemies in different parts of the world intended to depict this absolute victory, this definitive victory, as a doubtful victory."[36]

Opponents of the election outcome have pledged to stage new demonstrations on Saturday, raising concerns about possible government responses if Khamenei's demand that the protests end is not met. Many observers are beginning to frame the current stand-off in terms of a test of resolve, questioning how far the regime is willing to go to quell public unrest and how much Musavi and his supporters are willing to endure in order to see that their demands are met.

International Response

World wide, attention has focused on the events unfolding in Iran. Demonstrations took place in Western Europe and in other regions to protest the election outcome or the use of force against the demonstrators in Iran. Iranian expatriates also joined the protests. In the United Arab Emirates, protesters gathered in front of the Iranian consulate in Dubai to protest alleged election abuses. The consulate denied that any protests had taken place. After protests were held for five consecutive days, UAE authorities ordered the protestors to disperse.[37]

While some leaders offered congratulations to Ahmadinejad after the election,[38] others have withheld their felicitations until the Guardian Council's election probe is completed. The international community, particularly the United States and the European Union, has now focused its attention on the public demonstrations in Iran, but most official statements have been cautious, likely to avoid the appearance of interference in Iran's domestic affairs. The European Union expressed concerns about the alleged irregularities, adding that it "expects the new Government of the Islamic Republic of Iran will take its responsibility towards international community and respect its international obligations."

Britain's Foreign Secretary David Miliband also expressed concerns about the elections in Iran, saying that Britain had "followed carefully, and admired, the passion and debate" during the election and that the reports of irregularities and accusations of fraud were "a matter for the Iranian authorities to address" and that "our priority is that Iran engages with the concerns of the world community, above all on the issue of nuclear proliferation."[39]

U.S. Response

The Obama Administration's response has been cautious. President Obama expressed "serious concern" about the events in Iran and the allegations of election fraud. He also indicated after the election that he would pursue his policy of engaging with Iran to find a solution to the nuclear issue regardless of the outcome of the election. Some analysts fear, however, that recent events have diminished the prospects for diplomacy, particularly as use of the Basij to violently confront civilian protestors renewed concerns about Khamenei's disregard for human rights and basic civil liberties.

At a State Department press conference on June 17, U.S. Secretary of State Hillary Clinton said that the people of Iran deserve to have their voices heard and votes counted, and reiterated the position of other Administration officials that it is for the Iranians to determine how best to resolve the current situation in Iran. She also expressed the Administration's intent to pursue engagement regardless of the election outcome.

The U.S. government's response has been praised by some who argue that avoiding any appearance of involvement or meddling in Iranian affairs is the most likely choice to avoid provoking a harsh response from the Iranian government, one that would likely further endanger the lives of the demonstrators. These arguments tend to highlight the nationalist tendencies of Iranians from all parts of the political spectrum, particularly with regard to the complex history of intervention by the United States and other powers in Iran's domestic affairs. Others have criticized President Obama's response as too conciliatory toward Iran's existing leadership, particularly what some view as a lack of a strong enough condemnation of the use of force against civilians.

On June 16, 2009, President Obama drew criticism after saying in an interview with CNBC and the *New York Times* that, from an American national security perspective, there may not be a significant difference between Ahmadinejad and Musavi. He went on to say that the United States is going to be dealing with an Iranian regime that has historically been hostile to the United States. Critics of the statement argue that the President may be viewing the events in Iran solely through the lens of the nuclear issue.[40] Other analysts have argued that if Obama does not offer a stronger statement on the current situation it may be perceived by Iranians as a green light for Khamenei and the IRGC to use force to dispel the demonstrations.[41]

On June 19, the U.S. House of Representatives passed 405-1 H.Res. 560, which expresses support for Iranian citizens and "condemns the ongoing violence against demonstrators by the Government of Iran and pro-government militias, as well as the ongoing government suppression of independent electronic communication." The Senate also passed two measures— S.Res. 193 and S.Res. 196—which express support for Iranian citizens who "embrace freedom, human rights, civil liberties and rule of law" and which express the sense of the Senate on freedom of the press, freedom of speech, and freedom of expression in Iran.

Possible Outcomes and Implications for U.S. Policy

There is much debate about where the current situation in Iran could lead, with some experts predicting significant changes in Iran's political and social structure, and others arguing the Iranians at present do not appear to be seeking or experiencing a wholesale change in the basic nature of their government. Among those predicting significant change, some analysts are arguing a brewing "green revolution" is about to unfold in Iran. They cite the continued momentum of the protests, and say that the damage done by the regime to its own legitimacy is irreversible. Other observers have stated that the circumstances surrounding the election amount to a military coup, orchestrated by Khamenei and his allies in the military establishment. Both groups maintain that the current situation in Iran has caused or will cause significant changes that may even mean the end of the "Islamic Republic."[42]

At the opposite end of the spectrum are experts and practitioners who have argued that the public protests have more to do with Iranians' complaint that the Islamic Republic's electoral system was abused, rather than dissatisfaction with the notion of the Islamic Republic itself. These analysts tend to believe that some negotiated solution could possibly resolve the situation in Iran. Given the widespread popular dissatisfaction with the actions and statements of Supreme Leader Khamenei, it is possible that the doctrine of *velayet e faqih* that undergirds the supreme leader's position may be losing more support among some Iranians.

At this point, the short term outcome of the current stand-off is uncertain, but the determining factor could be the extent to which the state apparatus might be used to enforce what is, at the very least, widely perceived as an illegitimate election outcome.

Prospects for Engagement

The Obama Administration has maintained its commitment to engaging with Iran to resolve the issue of Iran's nuclear weapons program, its support for terrorism, and other national security concerns. Some analysts have speculated, however, that the outcome of the current stand-off in Iran may complicate or alter the course of U.S. policy toward Iran. Optimists consider the recent outpouring of public support for Musavi and calls for a new election from his supporters and some members of the clerical establishment as an indication that the Iranian public is no longer satisfied with the existing social contract, and may be less willing, as a result, to accept the international isolation that accompanies the government's position on the nuclear program, support for terrorism, and Ahmadinejad's rhetoric toward Israel. Even if the government manages to repress this popular opinion in the short run, some observers have acknowledged the possibility that, over the long run, the regime might not be sustainable against public unrest and widespread perceptions of illegitimacy.[43]

Skeptics see the recent events differently. Some argue that the engineered election outcome is a signal that Khamenei and the government establishment do not see engagement with the United States as a "prize to be won,"[44] and that no amount of diplomacy could change the perception that the United States is using the nuclear issue as a cover under which to pursue its real objective of regime change. As the Obama Administration works to strike a balance between not being perceived as interfering in Iranian affairs while appearing sympathetic to the civilian demonstrators, Khamenei has continued to accuse Western leaders of encouraging popular unrest. Continued calls for an end to the protests and warnings during his June 19 speech indicate, according to some analysts, that the Iranian government is prepared to take whatever means necessary to protect against a breakdown of the current system.[45]

Still another perspective is that the Iranian government might be willing to accept some negotiated solution to the current stand-off that would leave the regime intact, and perhaps even restore the civil-political balance. Some have argued that if the momentum of the protests continues to grow, then Khamenei and his allies in the government might sacrifice

Ahmadinejad in a factional realignment to preserve the regime. In this case, the course of U.S. policy might not change.[46]

End Notes

[1] Ali Akbar Hashemi Rafsanjani currently presides over the Assembly of Experts, a powerful body of clerics that has the authority to remove the supreme leader by a two-thirds vote.

[2] U.S. Open Source Center (OSC) Document IAP20090620l1002, "Iran: Text of Musavi's Letter to the Guardian Council," Tehran *Qalam* (in Persian), June 20, 2009.

[3] U.S. Open Source Center (OSC) Document IAP20090622950080, "Iran Guardian Council Spokesman: 'No Major Violations' in Vote Counting," *Vision of the Islamic Republic of Iran Network 1*, June 22, 2009.

[4] The President's Statement on Iran, June 20, 2009, available online at http://www.whitehouse.gov/blog/The-Presidents-Statement-on-Iran/.

[5] For more information and background on Iran, see CRS Report RL32048, *Iran: U.S. Concerns and Policy Responses*, by Kenneth Katzman.

[6] The Guardian Council is a 14-member body appointed by the supreme leader and responsible for overseeing elections, among other things.

[7] U.S. Open Source Center (OSC) Document IAP20090612950O1, "Editorial Says Entire Nation Will Win With Musavi Victory In Iran's Election," *E'temad Online*, June 10, 2009. See also Robert F. Worth, "In Iran, Harsh Talk as Election Nears," *New York Times*, June 8, 2009.

[8] Musavi's wife, Zahra Rahnavard, a political scientist and former chancellor of Alzahra University in Tehran, played a visible role in his campaign and garnered much support from female voters, an unconventional role in Iranian politics as wives of candidates have not in the past appeared at campaign events.

[9] U.S. Open Source Center (OSC) Document FEA20090610861590 "Iranian Election TV Debates Signal More Balanced Coverage," *BBC Monitoring*, June 10, 2009.

[10] See "Iranian Conservative Media Angry as Rafsanjani Lays Into Ahmadinejad," *BBC Monitoring*, June 10, 2009 and U.S. Open Source Center (OSC) Document IAP20099610950037, "Iranian President to Reply to Accusations Raised in Debates," *IRNA*, June 10, 2009. See also Thomas Erdbrink, "Ex-Iranian President Criticizes Ayatollah," *New York Times,* June 10, 2009.

[11]Robert F. Worth and Nazila Fathi, "Huge Campaign Rallies Snarl Tehran," *New York Times*, June 11, 2009.

[12]See Thomas Erdbrink, "Rural Iran May Shift Its Loyalty, *Washington Post*, June 7, 2009.

[13]See, for example, Marc Lynch, "Could There Be a Musavi Effect," *Foreign Policy*, June 10, 2009.

[14] U.S. Open Source Center (OSC) Document IAP20090611950041, "Iran: Candidate Musavi Warns Against IRGC, Basij 'Interference' in Election," *Farda* in Persian (Tehran), June 11, 2009.

[15] U.S. Open Source Center (OSC) Document IAP20090612950172, "Iranian Governors-General Authorized to Extend Voting Time," *Voice of the Islamic Republic Iran Radio* in Persian, June 12, 2009.

[16]U.S. Open Source Center (OSC) Document IAP2009061350119, *IRNA* (Tehran), June 13, 2009.

[17] U.S. Open Source Center (OSC) Document IAP2009061350138, "Leader Hails Iranians' Massive Turnout in Election," *Fars News Agency* (Tehran), June 13, 2009.

[18] For information on Iran's election law and procedures, see the International Foundation for Election Systems Election Guide for Iran, available at: http://www.electionguide.org/country-news.php?ID=103.

[19] U.S. Open Source Center (OSC) Document IAP20090615950012, "Musavi Files Formal Complaint Against Result in Iran's Presidential Election," *Fars News Agency* (Tehran), June 14, 2009.

[20] U.S. Open Source Center (OSC) Document IAP20090615950012, "Musavi Files Formal Complaint Against Result in Iran's Presidential Election," *Fars News Agency* (Tehran), June 14, 2009.

[21] See Eric Hooglund, "Iran's Rural Vote and Election Fraud," Tehran Bureau, June 17, 2009, available online at: http://tehranbureau.com/2009/06/17/irans-rural-vote-and-election-fraud/.

[22] U.S. Open Source Center (OSC) Document IAP20090615950044, "Iran's Supreme Leader Says Musavi's Letter to be Given Careful Consideration," *Islamic Republic of Iran News Network Television (IRINN)*, June 15, 2009.

[23] U.S. Open Source Center (OSC) Document GMP20090614647001, "Musavi Demands Elections Cancelled; Karrubi Rejects Ahmadinezhad's Win," *Al-Arabiyah* Television (Dubai), June 13, 2009.

[24] "Iran Probes 646 Poll Complaints," *BBC News*, June 18, 2009.

[25]See George Friedman, "Western Misconceptions Meet Iranian Reality," Stratfor Geopolitical Weekly, June 15, 2009.

[26] See, for example, Suzanne Maloney, "Reacting to Iran's Disputed Presidential Election Outcome," Brookings.edu, June 15, 2009, and Mehdi Khalaji, Patrick Clawson, Michael Singh, and Mohsen Sazegara, "Iran's 'Election': What Happened? What Does It Mean?, Washington Institute for Near East Policy, Policy Watch #1537: Special Policy Forum Report, June 18, 2009.

[27] U.S. Open Source Center (OSC) Document IAP20090613950098, " Iran's Interior Minister Warns Against Public Gatherings Without Permits," *Tehran Islamic Republic of Iran News Network Television (IRINN)*, June 13, 2009.

[28] "Iran Opposition Keeps Up Pressure," *BBC News*, June 16, 2009.

[29] David Clark Scott, "Iran's pro-Ahmadinejad media: Using Fake Crowd Photos?" *International News Editor*, June 17, 2009.

[30]Neil MacFarquhar, "Iran's Latest Protests are Seen as Toughest to Stop," *New York Times*, June 16, 2009.

[31]Robert F. Worth and Nazila Fathi, " Unrest Deepens as Critics are Detained," *New York Times*, June 15, 2009.

[32]Ian Black and Saeed Kamali Dehghan, "Iran Uprising Turns Bloody," *The Guardian* (UK), June 16, 2009.

[33] U.S. Open Source Center (OSC) Document, "New Protests Over Iran Elections, Public Media Revolt," *Kuwait KUNA Online*, June 17, 2009 and *BBC News*, June 17, 2009.

[34] "Masses Mourn Protesters in Iran," *BBC News*, June 18, 2009.

[35] "Khatami's Clerical Group Urges Annulling Iran Poll Results," *Associated Foreign Press*, June 13, 2009.

[36] "Ayatollah Demands End to Protests," *BBC News*, June 19, 2009 and U.S. Open Source Center (OSC) Feature FEA20090619863000, "Iran: Leader Khamene'i Calls for End to Protests," *Islamic Republic of Iran News Netwrok Television (IRINN),* June 19, 2009.

[37]Nour Sumaha, "Dubai Bans Iran Protests," *The National* (Abu Dhabi), June 17, 2009.

[38] Among those offering congratulations were the leaders of Syria, Lebanon, Indonesia, the Palestinian Authority, Turkey, Afghanistan, Russia, China, Iraq, and Head of Supreme Iraqi Islamic Council Abdul Aziz al-Hakim.

[39]U.S. Open Source Center Document EUP20090614086006, "Britain's Miliband Comments on Iranian Election," *London Foreign and Commonwealth Office*, June 13, 2009.

[40] Helene Cooper and Mark Landler, "For Obama, Pressure to Strike Firmer Tone," *New York Times*, June 18, 2009. Karim Sadjadpour of the Carnegie Endowment for International Peace added that "up until now, the president had very thoughtfully calibrated his marks on Iran," but called this particular statement an "uncharacteristic and egregious error." Sadjadpour and others have expressed concerns that such statements make Obama appear unsympathetic to the Iranians who are risking their lives to protest the elections by saying that the outcome does not matter to the United States.

[41] See Mehdi Khalaji, Patrick Clawson, Michael Singh, and Mohsen Sazegara, "Iran's 'Election': What Happened? What Does It Mean?", Washington Institute for Near East Policy, Policy Watch #1537: Special Policy Forum Report, June 18, 2009.

[42] Ibid. See also Mohsen Makhmalbaf interview with Foreign Policy, June 2009. Available online at http://www.foreignpolicy.com/story/cms.php?story_id=5018.

[43] See, for example, "U.S. Should React Cautiously to Iran's "Stolen Election," CFR Interview with Gary Sick, June 14, 2009 and Suzanne Maloney, "Reacting to Iran's Disputed Presidential Election Outcome."

[44] Mehdi Khalaji, Patrick Clawson, Michael Singh, and Mohsen Sazegara, "Iran's 'Election': What Happened? What Does It Mean?, Washington Institute for Near East Policy, Policy Watch #1537: Special Policy Forum Report, June 18, 2009.

[45]Ibid. See also, for example, Mehdi Khaliaji, "Khamene'i's Coup," *Washington Post*, June 15 ,2009

[46]See, for example, "Regime Adjusts Responses to Massive Protests," *Oxford Analytica*, June 16, 2009.

In: Iran Social, Economic and Political Developments
Editor: Joshua A. Owenstein
ISBN: 978-1-60741-552-7

Chapter 4

IRAN'S BALLISTIC MISSILE PROGRAMS: AN OVERVIEW*

Steven A. Hildreth

SUMMARY

Iran has an active interest in developing, acquiring, and deploying a broad range of ballistic missiles, as well as developing a space launch capability. This was spotlighted several times since 2008. In mid-July 2008, Iran launched a number of ballistic missiles during military exercises, reportedly including the medium-range Shahab-3. At the time, a Pentagon spokesman said Iran was "not testing new technologies or capabilities, but rather firing off old equipment in an attempt to intimidate their neighbors and escalate tension in the region." Subsequent analysis of the July 2008 missile launches shows Iran apparently digitally altered images of those launches. Iran announced other missile and space launch tests in August and November 2008. In February 2009, Iran announced it launched a satellite into orbit and "officially achieved a presence in space." This short report seeks to provide an overview of the reported or suspected variety of Iranian ballistic missile programs. Because there remains widespread public divergence over particulars, however, this chapter does not provide specificity to what Iran may or may not have, or is in the process of developing.

Iran's ballistic missile program dates to the late 1970s after the Shah was overthrown and the Islamic Republic of Iran established. The new Iranian government embarked on a ballistic missile program marked by considerable secrecy. Many consider that Iran's ballistic missile development was in full force by the mid-1980s during its protracted war against Iraq, during which Iran reportedly launched more than 600 ballistic missiles.

Today, there is little disagreement among most experts that Iran has acquired some number of ballistic missiles from other countries and has developed other ballistic missiles indigenously or in cooperation with others. Iranian ballistic missile proliferation has been a

---

* This is an edited, reformatted and augmented version of a CRS Report for Congress publication dated February 2009.

matter of U.S. and international concern. At the same time, however, there has been considerable public disagreement over precisely what kinds of ballistic missile systems Iran has or is developing itself or in cooperation with others. This is because there is little transparency in Iran's ballistic missile programs, which has led to some degree of a lack of confidence in Iran's public assertions of its activities. Finally, details about Iranian ballistic missile programs remain classified in the United States. Because of the secrecy inherent in the development of weapon systems, especially in less open societies, open-source analyses reflect a wide-range of technical views and assessments.

This chapter provides a brief description of what is publicly discussed regarding Iran's ballistic missile programs[1]; it does not discuss Iranian cruise missiles[2] or rockets.[3] These latter weapons were a source of concern when some reportedly Iranian-made rockets and other missiles may have been used by Hezbollah against Israel in 2006.[4] Charges of Iranian military support to Iraqi insurgents do not include Iranian-made rockets or missiles, however. This chapter first examines Iran's long-range ballistic missile programs because those efforts generally drive the greatest concerns within the United States, especially when coupled with concern over the development of Iran's nuclear capabilities. A brief overview of Iran's medium and short-range ballistic missile programs then follows.

Long-Range Ballistic Missiles

Traditionally, the United States has defined long-range or Intercontinental Ballistic Missiles (ICBMs) as those ballistic missiles capable of ranges greater than 5,500 kilometers (about 3,400 miles). To date, five countries have deployed operational ICBMs (all with nuclear weapons): the United States, Russia, China, France, and Britain. Other countries, such as Iran, are believed by some observers to have ICBM programs in varying stages of development.

In 1999, the U.S. intelligence community assessed that at some point the United States would probably face ICBM threats from Iran.[5] This remains the official U.S. position, that "Iran *could test* an ICBM in the last half of the next decade using Russian technology and assistance" (emphasis in the original). A similar report was issued in 2001.[6] This assessment is often interpreted that Iran will have ICBMs by 2015, but the unclassified intelligence statements place various caveats on that potential capability.

These intelligence statements serve as the official U.S. basis for assessing the Iranian ICBM threat to the United States and its friends and allies. These assessments drive U.S. military efforts designed to respond to such threats, such as the U.S. Ballistic Missile Defense (BMD) program in general and the U.S. proposed missile defense system in Europe specifically, as well as U.S. diplomatic efforts to curb Iranian long-range ballistic missile programs.[7] These assessments, in conjunction with official U.S. assessments of Iranian nuclear weapons development, contribute significantly to ongoing U.S. concerns over Iranian threats to U.S. and international security.

These assessments do not mean, however, that there is universal agreement within the U.S. intelligence community on the issue of an Iranian ICBM. According to these unclassified statements, some argue that an Iranian ICBM test is likely before 2010, and very likely before 2015. Other U.S. officials believe, however, that there is "less than an even chance" for such a test before 2015. Furthermore, U.S. assessments are also conditional in that an Iranian ICBM capability would have to rely on access to foreign technology, from, for example, North Korea or Russia.[8] Finally, it is argued that an Iranian ICBM could develop from an Iranian space program under which a space-launch vehicle program might be converted into an ICBM program. Some have argued that Iran could develop and test such a space launch vehicle by 2010.

Some observers argue that although the U.S. position may be based upon a realistic assessment, it is also a worst-case analysis of the potential threat from Iran. They argue that "with rare exception this level of threat has rarely turned out to be the historical reality."[9]

Beyond these general U.S. public statements about Iranian ICBM developments, there are few unclassified details. Further, non-official public sources reflect little technical or program consensus regarding an Iranian ICBM program. Some have referred to a program called the Shahab-6 (or Kosar in some instances) as a potential ICBM development program, perhaps derived from North Korean or Russian missile technology, or both.[10] Although Iran continues to declare it has no plans to develop an ICBM program, there appears to be considerable public uncertainty and debate as to whether the Shahab-6 is an actual design study concept, or an active or abandoned Iranian ICBM or even space-launch program.

In January 2004, Iran's Defense Minister reportedly announced that Iran would launch a satellite within 18 months. Iran then launched its first commercial satellite on a Russian rocket in 2005, and announced that it had allocated $500 million for space projects over the next five years. In February 2008, Iran reportedly launched a low-orbit research rocket in preparation for a later satellite launch. The Bush White House called that launch "unfortunate." In August 2008, Iran said it successfully launched a rocket that could carry its first satellite, but U.S. defense officials said the test aimed failed shortly after launch. In early February 2009, Iran successfully launched its Omid (Hope) satellite[11] on a Safir-2 (or Ambassador-2) rocket , which has a range of about 155 miles. A Pentagon spokesman said this launch was "clearly a concern of ours" because "there are dual-use capabilities here which could be applied toward the development of long-range missiles."

MEDIUM AND INTERMEDIATE-RANGE BALLISTIC MISSILES

Many experts believe that Iran's Shahab-3, which sometimes appears also to be called the Zelzal3 ballistic missile, is a derivative of the North Korean No-Dong 1 ballistic missile. It has a reported range of about 1,000 to 1,500 kilometers. This could reach potential targets throughout most of the Middle East. Some have speculated that North Korea, Iran, and Pakistan entered into a cooperative effort at one point to develop a missile of this range and capability. Other observers have alleged Russian assistance in Iranian development of this missile. Some reports suggest that Iran has already deployed a number of medium-range ballistic missiles (MRBMs). Of all Iranian ballistic missile programs, there seems to be more publicly available information in relative terms about this particular missile system than others.[12] Even so, there remains considerable and varying differences in open sources about this system.

Longer range versions of the Shahab-3, variously referred to as Shahab-3 variants, the Shahab3A, Shahab-3B, and Shahab-4, and a BM-25, may have range capabilities of 1,500-

2,500 kilometers. These missiles potentially could reach targets throughout the Middle East, Turkey, and into southeastern Europe. Some have reported that perhaps several dozen or more of these missile types may be deployed and operational. Some Chinese, North Korean, or Russian involvement is suspected. In 2006, Iran announced the successful test of a Fajr-3 MRBM comparable to the Shahab-3, although U.S. and Israeli intelligence analysts reportedly expressed skepticism.

In mid-July 2008, Iran launched a number of ballistic missiles and rockets of varying ranges during military exercises. Some observers noted the missile launches were in direct response to long-range Israeli air force exercises at the time. Iran claimed it flight tested a 2,000 km version of the Shahab-3 that could carry a 1-ton warhead. If accurate, this missile could hit targets throughout the Middle East and Turkey. But some analysts are skeptical of these Iranian claims pending further technical analyses of these recent military exercises, and others have cited Iranian exaggerations of its missile capabilities in the past. Various major international media retracted initial images of the missile launches because they were reportedly digitally altered. Bush Administration officials said Iran did not test new technologies or capabilities, but said the missile launches were evidence of the need for its proposed missile defense system in Europe. Secretary of State Rice stated the exercises were not helpful and that Iran should refrain from such activities.

Iran said it successfully test fired a 2-stage solid-fuel missile with a 2,000 kilometer range in November 2008. At the time, a Pentagon spokesman said he could not confirm the launch occurred, but that this was consistent with the fact that Iran continues to develop a ballistic missile program that poses a threat to Iran's neighbors in the region and beyond.

Other reports have also surfaced over Iran's development of a much longer MRBM with ranges of 4,000-5,000 kilometers, or even a space launch vehicle derived from these efforts that some refer to as the Shahab-5. The degree to which this effort might be actually underway also is highly uncertain.

Short-Range Ballistic Missiles

Iran is widely believed to have deployed a number of short-range ballistic missiles (SRBMs)—those with ranges less than 1,000 kilometers. In addition, Iran is believed to have various other SRBMs under development, either indigenously or in varying degrees of cooperation with countries such as China, North Korea, or Russia.

Beyond these speculations, however, open source materials do not reflect a consensus over technical capabilities and performance. Additionally, there appear to be considerable differences in descriptions of the numbers of systems operational or deployed and even the agreed-upon names of SRBMs ascribed to Iran. Some of the more commonly referred to Iranian missiles are discussed briefly.

Some believe that Iran may have imported perhaps 200 Chinese CSS-8 (or Tondar-69) SRBMs in the late 1980s, as well as a number of associated launch systems for their operational deployment.[13] The CSS-8 may have a range of about 150 kilometers.

Iran may have developed an SRBM in the 1990s called the Fateh A-110 (also apparently referred to as the Mershad or Zelzal-2 variant). According to various reports, this missile may have been developed with Chinese, Syrian, and North Korean involvement. This missile may have a range of about 200 kilometers and may have become operational around 2004.

Some believe that Iran acquired several dozen Chinese M-11 or CSS-7 SRBMs and associated launch vehicles in the mid-1990s, although China has denied this. The M-11 reportedly has a range of around 280 kilometers.[14]

Iran may also possess a number of SRBMs with ranges of 200-300 kilometers that it might have acquired from Libya or North Korea. Some may have been produced or modified indigenously. These have variously been referred to as the SCUD-B, SCUD-B variants, or Shahab-1 SRBMs.

Iran might also possess a few hundred SRBMs with a range of about 500-700 kilometers or so. These SRBMs have sometimes been referred to as the SCUD-C and Shahab-2. Analysts have expressed uncertainty over whether the Iranians developed and built these missiles on their own, or had help from China and North Korea.

Finally, there are some reports of an operational SRBM with a range up to 800 kilometers, which may possibly be referred to as an M-9 variant, DF-15, or CSS-6. Reportedly, the PRC produced the M-9 for export and Iran has acquired some number of them.

End Notes

1 After an initial powered phase of flight, a ballistic missile leaves the atmosphere (about 100 kilometers) and follows an unpowered trajectory or flight path before reentering the atmosphere toward a predetermined target. Ballistic missile ranges can vary from a hundred or so kilometers to more than 10,000 kilometers.

2 A cruise missile is a guided missile that normally uses some form of jet propulsion system to allow sustained flight within the atmosphere toward its target. Cruise missile ranges can vary from a few hundred to more than 1,500 kilometers.

3 After an initial powered launch, a military rocket will head toward its intended target without leaving the atmosphere. Military rocket ranges are relatively short range and can be guided or unguided.

4 For instance, see CRS Report RL32048, *Iran: U.S. Concerns and Policy Responses*, by Kenneth Katzman, and "The Rocket Campaign Against Israel During the 2006 Lebanon War," by Uzi Rubin, *Mideast Security and Policy Studies*, no. 71, June 2007. There remains uncertainty over the extent of Hezbollah's rocket and missile inventory and what was actually fired during the 2006 war. This includes the Chinese-built anti-ship missile (C-802), which Iran imported in the early 1990s. At least one such C-802 may have been transferred to Hezbollah and used to nearly sink an Israeli naval corvette ship in July 2006. Reports of a "Fajr-3" Iranian-made artillery rocket used in the 2006 war are sometimes erroneously confused with an Iranian medium-range ballistic missile.

5 National Intelligence Council, "Foreign Missile Developments and the Ballistic Missile Threat to the United States through 2015," September 1999. This unclassified National Intelligence Estimate was provided in open testimony to the Senate Foreign Relations Committee on September 16, 1999, by Robert D. Walpole, National Intelligence Officer for Strategic and Nuclear Programs, Central Intelligence Agency.

6 National Intelligence Council, "Foreign Ballistic Missile Threat Through 2015," December 2001, Unclassified Summary of a National Intelligence Estimate.

7 See CRS Report RL34051, *Long-Range Ballistic Missile Defense in Europe*, by Steven A. Hildreth and Carl Ek, and CRS Report CRS Report RL32048, *Iran: U.S. Concerns and Policy Responses*, by Kenneth Katzman.

8 See National Intelligence Council, "Foreign Missile Developments, 1999," and Robert D. Walpole, "The Ballistic Missile Threat to the United States," Statement Before the Senate Subcommittee on International Security, Proliferation and Federal Services, February 9, 2000.

9 For instance, see Charles P. Vick, "North Korean, Iranian, and Pakistani Common Russian, Chinese Nuclear Weapons Heritage and Tests, What does it Reveal about the Missile Borne Warhead Development Status?" Part 1, March 20, 2007, at http://www.globalsecurity.org/wmd/world/dprk/nuke-warhead-dev1.htm.

10 See, for instance, "Shahab-6 IRSL-X-4," at http://www.fas.org/nuke/guide/iran/missile/shahab-6.htm, and "Shahab6," at http://www.missilethreat.com/missilesoftheworld/id.110,css.print/missile_detail.asp.

[11]Space-Track.org provides access to orbital data received from the Department of Defense, and shows two orbiting Iranian objects launched Feb. 2, 2009. One is the satellite and the other is the final stage of the rocket.

[12] See, for instance, the Federation of American Scientists: http://www.fas.org/nuke/guide/iran/missile/shahab-3.htm and http://www.missilethreat.com/missilesoftheworld/id.107,css.print/missile_detail.asp.

[13] The International Institute for Strategic Studies, *The Military Balance 2007*, p. 224.

[14]http://missilethreat.com/missilesoftheworld/id.66,css.print/missile_detail.asp.

In: Iran Social, Economic and Political Developments
Editor: Joshua A. Owenstein
ISBN: 978-1-60741-552-7

Chapter 5

IRAN'S ECONOMIC CONDITIONS: U.S. POLICY ISSUES*

Shayerah Ilias

SUMMARY

The Islamic Republic of Iran, a resource-rich and labor-rich country in the Middle East, is a central focus of U.S. national security policy. The United States asserts that Iran is a state sponsor of terrorism and that Iran's uranium enrichment activities are for the development of nuclear weapons. To the extent that U.S. sanctions and other efforts to change Iranian state policy target aspects of Iran's economy as a means of influence, it is important to evaluate Iran's economic structure, strengths, and vulnerabilities.

Since 2000, Iran has enjoyed broad-based economic growth. However, strong economic performance has been hindered by high levels of inflation and unemployment and low levels of foreign investment. Some contend that President Ahmadinejad's expansionary monetary and fiscal policies have worsened unemployment, inflation, and poverty in Iran. Iran's economic growth is expected to slow in 2009.

Iran has long been subject to U.S. economic sanctions, and more recently, to United Nations sanctions, over its uranium enrichment program and purported support for terror activities. Such sanctions are believed by some analysts to contribute to Iran's growing international trade and financial isolation.

Iran's economy is highly dependent on the production and export of crude oil to finance government spending, and consequently is vulnerable to fluctuations in international oil prices. Although Iran has vast petroleum reserves, the country lacks adequate refining capacity and imports gasoline to meet domestic energy needs. Iran is seeking foreign investment to develop its petroleum sector. While some deals have been finalized, reputational and financial risks may have limited other foreign companies' willingness to finalize deals.

---

* This is an edited, reformatted and augmented version of a CRS Report for Congress publication dated June 2009.

While Iran-U.S. economic relations are limited, the United States has a key interest in Iran's relations with other countries. As some European countries have curbed trade and investment dealings with Iran, other countries, such as China and Russia, have emerged as increasingly important economic partners. Iran also has focused more heavily on regional trade opportunities, such as with the United Arab Emirates.

High oil prices have increased Iran's leverage in dealing with international issues, but the country's dependence on oil and other weak spots in the economy have to come to light by the 2008 international financial crisis, which may portend a slowing down of Iran's economy.

Members of Congress are divided about the proper course of action in respect to Iran. Some advocate a hard line, while others contend that sanctions are ineffective at promoting policy change in Iran and hurt the U.S. economy. In the 110th Congress, several bills were introduced that reflect both perspectives. Policies toward Iran remain a key issue for the 111th Congress.

INTRODUCTION

The Islamic Republic of Iran is a central focus of U.S. national security policy. The United States has designated the Iranian government as a state sponsor of terrorism. The United States contends that Iran is a destabilizing force in the Middle East and expresses concern about its growing influence in the region and internationally. The United States has decried Iran's uranium enrichment activities, which allegedly are being used to develop nuclear weapons. Iran also has been accused of arming Shiite militias in Iraq, providing support to Hezbollah and Hamas, and inflaming sectarian strife in the Middle East.

This chapter provides a general overview of Iran's economy, addresses related U.S. policy concerns, and discusses policy options for Congress. The purpose of this chapter is two-fold. First, it provides insight into important macroeconomic trends, policy reforms and objectives, key economic sectors, international trade patterns, and sources of foreign exchange.[1] Second, in the context of U.S. economic sanctions imposed for national security and foreign policy reasons, the report evaluates Iran's economic structure, strengths, and vulnerabilities and discusses issues and options for Congress.

Iran boasts the word's third largest petroleum reserves, following Saudi Arabia and Canada, and the second largest gas reserves, after Russia. Iran also has the Middle East and North Africa region's second largest economy, after Saudi Arabia, and the second largest population, after Egypt (see **Table 1**). Nevertheless, Iran faces a number of significant economic challenges. Internal challenges include dependence on oil export revenues to finance government spending and vulnerability to oil price fluctuations; dependence on gasoline imports to meet domestic energy needs; high inflation, unemployment levels, and poverty levels; reported domestic economic mismanagement; and widespread economic inefficiency. External challenges include U.S. and United Nations (U.N.) sanctions and other forms of U.S.-led financial pressure and the fallout from the recent global economic turndown.

Table 1. Iran Country Overview

Indicator	Description
Land Area	1.6 million square kilometers (slightly larger than Alaska) (CIA)
Population	65.4 million (CIA, July 2009 estimate)
Median Age	26.4 years (CIA, July 2008 estimate)
Head of State	Mahmoud Ahmadinejad, President since August 2005
Capital	Tehran
Life Expectancy at Birth	71.14 years (CIA, 2009 estimate)
Gross Domestic Product (GDP)	$768 billion (purchasing power parity) (EIU FY2007);
	$724 billion (real GDP translated into U.S. dollars using PPP exchange rate in 2005) (EIU, FY2007 estimate)
GDP Real Growth Rate	7.8% (EIU, FY2007 estimate)
GDP Per Capita	$10,781 (EIU, FY2007 estimate)
GDP Composition by Sector (Current Prices)	Industry, 44.3%; services, 44.9%; agriculture, 10.8% (CIA, 2008 estimate)
Unemployment Rate	12.5%, reported by Iranian government (CIA, 2008 estimate)
Population Below Poverty Line	18% (CIA, 2007 estimate)
Inflation Rate (Consumer Prices)	17.1% (EIU, FY2007)
Exports	$91 billion (IMF, FY2007)
Export Commodities	Petroleum, chemical and petrochemical products, fruits and nuts, carpets (CIA)
Imports	$55 billion (IMF, FY2007)
Import Commodities	Industrial raw materials and intermediate goods, capital goods, foodstuff and other consumer goods, technical services (CIA)

Sources: Central Intelligence Agency (CIA) Factbook; IMF, Direction of Trade Statistics
Note: The Iranian fiscal year runs from March 21st to March 20th.

HISTORICAL CONTEXT

The 1979 Islamic revolution changed Iran's modern political and economic history. Ayatollah Ruhollah Khomeini and his supporters transformed Iran into an Islamic state with a public sector-dominated economy that was increasingly internationally isolated. With the Iran-Iraq war (1980-1988), Iran faced negative rates of real economic growth, declines in oil production and revenue, and high levels of inflation. This represented a reversal of economic prosperity in 1960s and 1970s, during which Iran's economy experienced real economic growth rates nearing 10%, one of the world's highest, along with growth in per capita income and low inflation levels.[2]

During the 1990s, Iran strived to rebuild war-torn local production, attract international investment, enhance foreign relations, liberalize trade, and, more recently, redistribute wealth under a series of a five-year economic plans. Post-war economic growth included recovery in oil output, but the country faced a severe economic downturn in the latter part of the decade due to a drop in international oil prices.[3]

Iran is the second largest oil producer in the Organization of the Petroleum Exporting Countries (OPEC). The country's oil and gas reserves rank among the world's largest. Iran's economy is largely dependent on oil and is highly susceptible to oil price shocks.

Since the 1979 U.S. embassy hostage crisis in Tehran, Iran has been subject to various U.S. economic sanctions. Such actions have been motivated over time by concerns regarding Iran's nuclear program and support for terrorist organizations. More recently, the United States increasingly has focused on targeted financial measures to isolate Iran from the U.S. financial and commercial system. Sanctions have been imposed in order to change the Iranian government's policies with respect to its nuclear program and support to terrorist organizations. To that end, the United States has imposed sanctions to curtail the development of Iran's petroleum sector and constrain Iran's financial resources in a way that motivates policy change in Iran. The United States also has applied diplomatic pressure on foreign countries and companies to limit business with Iran.[4] In addition to the United States, some European Union states and other countries have imposed sanctions on Iran in line with moves by the United Nations (U.N.).

The United States has pushed for stronger international sanctions against Iran in the United Nations. Most recently, in March 2008, the United Nations Security Council (UNSC) passed a third round of sanctions against Iran through Resolution 1803, calling for the inspection of suspicious international shipping to and from Iran that are suspected of carrying prohibited goods. It encourages greater monitoring of named Iranian financial institutions, travel bans for named Iranians, and freezes of additional assets related to Iran's nuclear program.

In June 2008, the five permanent members of the UNSC (Britain, China, France, Russia, and the United States) and Germany offered to suspend further sanctions against Iran if Iran agreed to halt its uranium enrichment program and to begin negotiations on constraints of its nuclear activity.[5] The six countries considered Iran's response unclear, and in August 2008, they agreed to pursue a fourth round of U.N. sanctions against Iran.[6]

Iran has opposed U.S. and U.N. sanctions vehemently. The country has long maintained that the purpose of its uranium enrichment program is to produce fuel for nuclear power reactors, rather than fissile material for nuclear weapons. The Iranian government asserts its right to develop nuclear energy for peaceful purposes. Iran increasingly has questioned the justification of the sanctions in light of some recent positive reports on its nuclear activities. A November 2007 U.S. National Intelligence Estimate (NIE) assessed that Iran stopped its nuclear activities for weapons proliferation in 2003.[7] Iran and the International Atomic Energy Agency (IAEA) agreed in August 2007 on a work program that would clarify the outstanding questions regarding Tehran's nuclear program. Iran has clarified some questions, but a May 2008 report by the IAEA raised major new questions about Iran's nuclear intentions.[8]

Table 2. Iran's Average Annual Real GDP Growth: 2000-2010

Fiscal Year	Average Annual Growth (%)
2000	5.1
2001	3.7
2002	7.5
2003	7.1
2004	5.1
2005	4.7
2006	5.8
2007	7.8
2008 (estimate)	6.5
2009 (forecast)	0.5
2010 (forecast)	2.9

Source: Economic Intelligence Unit (derived from World Bank, World Development Indicators).

Overview of Iran's Economy

Since FY2000, Iran's economy has experienced real economic growth rates of about 6.4% on average annually.[9] The annual change in real GDP reached a high of 7.8% in FY2007 (see Table 2). Recent economic growth has been driven by government spending on priority sectors, expansionary monetary and fiscal economic policies, increased growth in credit, and private consumption. Despite high international oil prices in recent years, the contribution of the oil and gas sector to economic growth has been more modest. The oil economy has been faced with low levels of production and inadequate investment.[10] U.S. and U.N. sanctions levied against Iran, along with the poor domestic business environment, may contribute to low levels of investment. Iran's economic growth is expected to slow in 2009 and 2010, owing to the decline in international oil prices in late 2008, domestic economic mismanagement, and limited oil revenue savings to weather the recent global economic turndown.[11]

Iran's economic growth has been hampered by consistently double-digit rates of inflation. Although high inflation is widespread among the oil-exporting countries in the Middle East and Central Asia, Iran has one of the highest.[12] Iran's average Consumer Price Index (CPI) inflation reached 17.1% in FY2007 and is projected to grow to 28.0% for FY2008.[13] By some estimates, Iran's inflation level is over 30% presently.[14] Iranians struggle with the rising cost of basic foods, such as rice, chicken, and eggs, and housing prices,[15] which have eroded real wages. The poor are hit hardest by inflation. It is the poor, mainly from rural areas, who supported President Ahmadinejad in the 2005 presidential election. Support for Ahmadinejad weakened marginally during the March 14, 2008 parliamentary elections, despite Iran's economic difficulties.

Domestic factors contributing to the rise in inflation include expansionary government economic policies and growing consumption demands. External factors include international sanctions against Iran and rising international food and energy import prices.[16] Inflation levels have been associated with Ahmadinejad's efforts to curb banking interest rates for loans to sub-inflation levels. The Central Bank has opposed these hikes. Inflation levels are expected to ease in the coming months due to the decline in international prices for oil and other commodities.

The unemployment rate remains high, reaching an estimated 12.5% in 2008.[17] Some observers contend that the unemployment rate is higher than figures reported by the Iranian government. At least one-fifth of Iranians lived below the poverty line in 2002.[18] Iran has a young population[19] and each year, about 750,000 Iranians enter the labor market for the first time, placing pressure on the government to generate new jobs.[20] The emigration of young skilled and educated people continues to pose a problem for Iran. The IMF reported that Iran has the highest "brain drain" rate in the world.[21]

Economic Policy and Reform Efforts

Over the past few decades, Iran has engaged in a series of five-year economic plans in order to shift its state-dominated economy into an economy that is market-oriented, private sector-led, and economically diversified. Reform efforts have experienced resistance from various elements of Iran's political establishment.

Significant strides toward trade liberalization, economic diversification, and privatization took place under the Khatemi administration (1997-2005). The government introduced some structural reforms such as tax policy changes and adoption of new foreign investment laws to promote Iran's global market integration and attract investment. Iran shifted to a unified managed float exchange rate system in March 2002.[22] At various times previously, Iran has had different combinations of exchange rates, including official, export, parallel market, and Tehran stock market versions. The exchange rate reform is considered to have improved Iran's trading environment and to have enhanced public sector transparency modestly.[23]

President Ahmadinejad has taken a more populist approach with his economic policies, with promises of "bringing the oil money to people's tables" when he took office in 2005. Some critics maintain that policies under President Ahmadinejad have been a major contributor to budget deficits and are ineffective tools for combating inflation, unemployment, and poverty.[24]

In line with Ahmadinejad's populist agenda, fiscal policy has been expansionary. The government provides extensive public subsidies on gasoline, food, and housing. Energy subsidies alone represent about 12% of Iran's GDP. Some observers estimate total subsidies to reach over 25% of GDP. When including implicit subsidies, the government's spending on subsidies may be even higher. In addition to subsidies, President Ahmadinejad has provided cash handouts to the poor. Subsidies and cash handouts are considered by many to be un-targeted and ineffective at helping the poor. The IMF has encouraged Iran to reduce its subsidies.[25]

Monetary policy also has been expansionary under Ahmadinejad. The government has provided low-interest loans for agriculture, tourism, and industry and has instituted loan forgiveness policies. Other activities include the creation of a number of social programs to assist farmer and rural residents.[26] Ahmadinejad's cabinet established the $1.3 billion Imam Reza Mehr Fund (Imam Reza Compassion Fund) to assist youth with marriage, housing, and education in 2006.[27] As in other Middle Eastern countries, the rising cost of marriage is financially prohibitive to many young Iranians. Interest-free loans are available to youth for marriage through the fund. Some economists contend that Ahmadinejad's efforts to lower the interest rate have led to excessive liquidity and inflation.[28]

The government has used oil export revenues from the Oil Stabilization Fund (OSF) to support expansionary fiscal and monetary policies. The OSF was created by Iran's Central Bank, the Bank Markazi, in 2001 to store surplus oil revenue and to smooth economic vulnerabilities associated with oil price fluctuations.[29] Iran has been drawing down on its OSF to finance discretionary spending, such as for public subsidies, cash handouts to the poor, and low-interest loans. Of primary concern to the United States and the international community is the purported use of oil export revenues to finance Iran's nuclear program and alleged support for terrorist groups.

Economic concerns continued to erode President Ahmadinejad's political support base in the weeks before the March 2008 parliamentary elections.[30] Ahmadinejad's party appeared to emerge from the elections with significant continued support, but some allege that this is because many reformist candidates were disqualified from running. Others say that Ahmadinejad's faction successfully painted Iran's economic difficulties as caused by U.S. pressure.[31]

Analysts debate the extent to which Iran's economic policies are a result of poor decisions by the Iranian government and sub-optimal choices taken by the government in response to U.S. and international sanctions. Some economists believe that the sanctions augment the government's tendency toward state-led rather than private sector-oriented market policies.[32]

Iran and the Recent Global Economic Turndown

The recent global economic turndown has placed downward pressure on international oil prices. Given that oil exports account for the bulk of Iran's exports and government revenue, the economic downturn is expected to lead to slower economic growth, with annual growth in GDP expected to decline from an estimated 6.5% in FY2008 to a forecasted 3.5% in FY2009.[33] The economic downturn also may lead to higher levels of unemployment and a reduction in foreign exchange reserves for Iran. On the upside, lower oil prices may ease inflationary pressures. In a reversal of previous public stances, in December 2008, Ahmadinejad admitted for the first time that the decline in international oil prices will have a negative effect on Iran's economy.[34]

While estimates vary about the size of the Oil Stabilization Fund (OSF), many observers express concern that there are no longer sufficient funds in the OSF to cushion against the global economic turndown.[35] Ahmadinejad has been drawing down on the OSF to fund his expansionary economic policies and, reportedly, to fund Iran's uranium enrichment program and terror support. In light of the recent fluctuations in oil prices, the government may revise its budget estimates for next year. Previously, the government planned to base the budget on an oil price of $50 to $60 per barrel, but has considered using a lower oil price estimate.[36] Ahmadinejad proposed a plan to eliminate energy subsidies and to provide monthly allowances to the poor instead. Critics contended that the plan would push inflation higher and is an effort to boost Ahmadinejad's election-year support.[37] There are also concerns that elimination of government subsidies may lead to social instability.

Iran's isolation from the financial world due to U.S. and international sanctions has provided some insulation for Iranian banks and the Tehran Stock Exchange (TSE) from the global financial crisis. However, between October and December of 2008, the TSE fell by nearly 20%, with investor confidence shaken by the global economic turndown and the impact of declining oil prices on Iran's economy.[38] Moreover, as world credit markets dry up, Iran has faced decreased options in circumventing U.S. sanctions and accessing foreign investment.[39] Iran's weakened economic position may reduce its leverage against the U.S. and other foreign countries over its uranium enrichment program and reported terror financing.

ECONOMIC STAKEHOLDERS

Iran's economy is heavily dominated by the state, which is the recipient of revenues from crude oil exports, and quasi-state actors, such as the *bonyads* and the commercial entities of the Islamic Revolutionary Guard Corp (IRGC). Private sector activity is limited, although the government is engaged in some privatization efforts.

Bonyads

Sometimes referred to as "Islamic conglomerates," *bonyads* (Persian for "foundation") are semi-private charitable Islamic foundations or trusts that are believed to wield enormous political and economic power in Iran. They were among the institutions used by the regime to help nationalize Iran's economy after the 1979 revolution. *Bonyads* are not subject to the Iranian government's checks and balances. *Bonyads* report directly to the Supreme Leader and are not subject to parliamentary supervision. They do not fall under Iran's General Accounting Law and, consequently, are not subject to financial audits. Because *bonyads* are not required to disclose their financial activities, it is not known exactly the magnitude of their wealth.

The largest Iranian charitable trust is the Foundation of the Oppressed and War Veterans (*Bonyad e-Mostazafan va Janbazan*, MJF). With more than 200,000 employees and 350 subsidiaries, the MJF has an estimated value of more than $3 billion, at least 10% of Iran's gross domestic budget (GDP). The MJF provides financial assistance, medical care, and recreational opportunities to Iran's poor and individuals wounded or disabled from the Iran-Iraq war. Through its company affiliates, the MJF is involved in both Iran's domestic economy and foreign economies. The MJF's domestic economic scope is expansive, with affiliates involved in economic areas such as agriculture, construction, industries, mining, transportation, commerce, and tourism. Since 1991, the MJF has invested in energy, business, engineering, and agricultural activities in Europe, Russia, Asia, the Middle East, and Africa. Some allege that the MJF is used for Iranian intelligence activities for buying dual-use products for proliferation of weapons of mass destruction (WMDs).[40]

Many believe that *bonyads* enjoy a significant advantage over private companies. Prior to the unification of Iran's exchange rate system, the *bonyads* were able to access foreign exchange at deep discounts compared to private enterprises. Presently, *bonyad* officials have longstanding connections with politicians, and frequently get special access to credit at state-owned banks.[41] In addition, *bonyads* get privileges on taxation and import duties. Some critics contend that economic and political reform in Iran will not be significant unless *bonyads* are reformed. Some also contend that they contribute to political corruption and limit the funneling of oil wealth to the poor. *Bonyads* also may limit privatization, because shares for many of Iran's national companies undergoing privatization are given to *bonyads*, rather than wholly private enterprises.

Islamic Revolutionary Guard Corps

The Islamic Revolutionary Guard Corps (IRGC) was founded in 1979 by the Ayatollah Khomeini and is a branch of the Iranian government's military. The IRGC is comprised of five branches: the Grounds Force, Air Force, Navy, Basij militia, and Qods Force special operations.[42]

The Islamic Revolutionary Guard Corps increasingly is becoming an important player in the Iranian economy. The IRGC's initial economic involvement consisted of postwar reconstruction activities, largely infrastructure projects. More recently, the IRGC has become involved in commercial activity in the construction, oil and gas, and telecommunications sector.[43] Elements of the Iranian private sector have expressed displeasure with the IRGC. Through its powerful connections, the IRGC frequently acquires business contracts for new projects at the expense of private sector businesses. The IRGC also serves as a leading investment tool for many of Iran's leaders.

Some Iranians express concern that the IRGC is involved in Iran's underground economy. The IRGC has significant control over Iran's borders and airports. The IRGC is allegedly involved in smuggling alcohol and other low-level contraband into Iran. Some report that the IRGC smuggles gasoline, which is heavily subsidized in Iran, to other countries for profit.[44]

Some analysts believe that the Revolutionary Guard benefits from Iran's economic isolation. With foreign businesses unwilling or unable to enter into deals, the Revolutionary Guard faces less competition for acquiring new contracts. However, because the IRGC frequently does not have the technical expertise that many international companies do, the IRGC sometimes subcontracts to international companies, making a profit as an intermediary in the transaction.[45]

The United States contends that the IRGC is involved in proliferation of weapons of mass destruction (WMDs). Under Executive Order (E.O.) 13382, the United States can sanction entities for proliferation concerns. The sanctions prohibit all transactions between U.S. persons and the sanctioned entity and freeze any assets that the sanctioned entity has in the United States.[46] On October 25, 2007, under E.O. 13382, the U.S. Department of State designated the IRGC for proliferation concerns. The U.S. embargo on the IRGC represented the first time that the United States has sanctioned a foreign country's military. Also on the same day and under the same executive order, the U.S. Treasury identified nine companies either owned or controlled by the IRGC and five individuals associated with the IRGC for proliferation concerns.[47] These companies all are reportedly tied to Iran's energy sector.[48] They are listed below:

Companies:

- Khatam al-Anbya Construction Headquarters: Main engineering headquarters of the IRGC; secured deals of at least $7 billion in oil, gas, transportation, and other sectors[49]; owned or controlled by the IRGC
- Oriental Oil Kish: Drilling company; owned or controlled by the IRGC
- Ghorb Nooh: Owned or controlled by the IRGC
- Sahel Consultant Engineering: Owned or controlled by the IRGC
- Sepasad Engineering Company: Owned or controlled by the IRGC
- Omran Sahel: Owned or controlled by the IRGC
- Hara Company: Engineering firm associated with Khatam al-Anbya; owned or controlled by the IRGC
- Gharagahe Sazandegi Ghaem: Business services company owned or controlled by the IRGC

Individuals:

- General Hosein Salimi: Commander of the Air Force, IRGC
- Brigadier General Morteza Rezaie: Deputy Commander, IRGC
- Vice Admiral Ali Akhbar Ahmadian: Former Chief of the IRGC Joint Staff
- Brigadier General Mohammad Hejazi: Former Commander of Bassij resistance force
- Brigadier General Qasem Soleimani: Commander of the Qods Force

In addition to WMD proliferation concerns, the United States asserts that the IRGC is involved in terrorist activities. E.O. 13224 permits the President to freeze the assets of terrorists and their supporters.[50] On October 25, 2007, the United States sanctioned the IRGC-Qods Force under E.O. 133224. The United States asserts that the Qods Force provides to Hezbollah's military and terrorist activities, with assistance ranging between $100 to $200 million a year.[51]

Private Sector

Prior to the 1979 revolution, Iran boasted a vibrant, significant private sector. However, under the leadership of the Ayatollah Khomeini, the bulk of private sector companies, including commercial banks, were taken over by state and quasi-state institutions. Foreign participation in Iran's economy was prohibited.

Currently, wholly private enterprises are present in agriculture, trade, small-scale manufacturing, and mining, but play a minimal role in large-scale economic activity.[52] In an effort toward more private sector development, Iran began a major privatization initiative in July 2006. It allowed issuances of up to 80% of shares in strategic industries through the stock market, including downstream oil sector businesses, banks, insurance, utilities, and transportation.[53] Iran is also working to privatize state-run oil and gas companies.

Iran's private sector competes with the businesses operated by the *bonyads* and the IRGC. For example, Iranian officials have encouraged foreign companies to enter into the Iranian market. However, many business contracts have been won by quasi-state actors, such as the *bonyads* and commercial entities of the IRGC. In addition, the private sector is critical of the government's use of assets in the OSF to fund state-run companies at the expense of loans to private businesses.[54]

Some observers are critical of the Iranian government's continued strong involvement in the country's economy. Some Iranians believe that the government needs to invest oil export revenues in Iran's private sector rather than spending revenues on imports[55] and socially minded programs.

Historically, Iran has been a society of trade merchants, the *bazaari* class. As manufacturing in Iran is limited, merchants import goods, mark up the goods for profit, and then sell. In order to be economically viable, the *bazaaris* need low employment costs, low rents, free trade, and low regulation. The *bazaaris* tend to be skeptical of a large government role in the economy. They are supportive of Iranian trade with foreign countries. However, they tend to be critical of foreign investment because it would open up their companies to foreign competition.[56]

ECONOMIC SECTORS

Iran's economy has a number of key sectors. In 2008, industry – which includes oil and gas, petrochemicals, steel, textile, and automotive manufacturing – accounted for an estimated 44% of the Iran's GDP. The services sector, including financial services, represented about 45% of Iran's economy. Agriculture constituted about 11% of Iran's economy.[57] Agriculture continues to be one of the economy's largest employers, representing one-fifth of all jobs based on a 1991 census.[58] Some analysts have expressed concern that excessive focus on the oil and gas sector is crowding out investment and expansion opportunities in other sectors and opportunities for economic diversification.[59]

Oil and Natural Gas

Holding an estimated 10% of global proven oil reserves, Iran boasts the world's third largest proven petroleum reserves following Saudi Arabia and Canada. Most of Iran's crude oil reserves are in the southwestern region near the Iraqi border. Among the Organization of the Petroleum Exporting Countries (OPEC) members, Iran is the second largest oil producer following Saudi Arabia. Presently, Iran produces about 4 million barrels of crude oil per day (mbd), approximately 5% of total global production. While oil export revenues have spiked in recent years due to a surge in oil prices, Iran's crude oil output has remained essentially flat.

Iran' oil production levels are limited by a number of factors. The oil industry faces a high rate of natural decline of mature oil fields and low oil recovery rates. It is believed that millions of barrels of oil are lost annually because of damage to reservoirs and these natural declines. Iran also has been plagued by aging infrastructure and old technology. Structural upgrades and access to new technologies, such as natural gas injections and other enhanced oil recovery efforts, have been limited by a lack of investment, due in part to U.S. sanctions.[60] The United States is restricted from oil development investments in Iran, but other countries, until recently, have actively invested in Iran's oil sector development.

With an estimated 15% of the world's gas reserves, Iran has the second largest natural gas reserves globally, following Russia. [61] Despite its vast gas resources, Iran has been unable to become a major international gas exporter. In fact, Iran was a net importer of natural gas as late as 2005. Natural gas production could be used for domestic consumption, exports to European and Asian markets, and development of Iran's petrochemicals industry.[62] Iran has been seeking international investment to help build its natural gas sector. However, U.S. sanctions have limited Iran's access to technologies from abroad that are necessary for developing liquefied natural gas plants. The intellectual property for these technologies belong to a small network of U.S. and Japanese companies. Providing such technologies to Iran would violate the U.S. trade ban on Iran.[63]

The oil and gas sector is heavily state-dominated. Iran is engaging in efforts to privatize nearly 50 state-run oil and gas companies, estimated to be worth $90 billion, by 2014 through the Tehran Stock Exchange. Both domestic and foreign investors would be able to buy shares. Privatization of these energy companies may make it easier for investors to circumvent U.S. sanctions, which complicate investors' ability to engage in business transactions with Iran directly.[64]

Agriculture

Iran is a major world provider source of caviar and pistachio nuts, a significant non-oil export for Iran. The country's climate and terrain also support tobacco, tea, wheat and barley, among other food commodities. Iran's agriculture production is vulnerable to periodic droughts, including a severe drought in 2008. In addition to climate change, the agricultural sector faced setbacks in production during the 1979 revolution and the war with Iraq.[65] Overfishing and environmental degradation also threaten the agriculture sector.

Iran typically has used oil export revenues to pay for agricultural imports. However, rising international food commodity prices combined with a large population increase have placed pressure on Iran's economy, despite high international oil prices. Other Middle Eastern countries are experiencing similar economic strains.[66]

Manufacturing

Iran is working to build up various industries within its manufacturing sector, including steel, automotives, food products, and petrochemicals. There is some concern that Iran's manufacturing sector has declined because oil export revenues have increased Iran's exchange rate, making the manufacturing sector less competitive.[67] Manufacturing activity reportedly has been impeded by international sanctions. Iranian manufacturing units rely on imported parts and services from Europe. Access to imported intermediate goods has been complicated because a number of European banks have scaled down financial transactions with Iranian businesses.[68]

Steel

Iran is the largest producer of steel in the Middle East[69] and a significant producer of steel globally. Despite Iran's high production levels, the country is a net importer of steel.[70] There has been a ramp-up of growth in demand for steel in the Middle East, fueled by the need for investments in energy project infrastructure and expansion of construction activity. In April 2009, a business forum between Iran and Germany discussed ways of increasing bilateral trade between the countries, focusing particularly on steel.

Automotives

In 2007, Iran was the 16th largest motor vehicle producer in the world and the largest automaker among the Middle Eastern countries. Iran produces both light and heavy vehicles.[71] Its two biggest automakers are Iran Khodro and Sapia.[72] Auto plants frequently have outdated technology and parts must be imported through third countries. Cars frequently are not fuel-efficient, contributing to pollution.[73] Despite Iran's high level of automotive production, domestic demand for motor vehicles exceeds supply. Iran imports a variety of vehicles, including basic models, luxury vehicles, and vehicles for construction and mining. Iran reduced the tariff rate on auto imports in 2006. Iran recently began joint ventures with foreign companies for auto production, including Peugeot and Citroen (France), Volkswagen (Germany), Nissan and Toyota (Japan), Kia Motors (South Korea), Proton (Malaysia), and Chery (China).[74] Foreign companies have entered the Iranian auto market with some caution in light of concerns about U.S. reaction and reputational risks.

Food Products

Additionally, there has been a growth in agriculture-related manufacturing, such as rice milling and manufacturing of canned food and concentrates, fruit juices, and confectionary. Foreign companies, such as Nestle, Coca Cola, and Pepsi have signed deals for production with local Iranian businesses.[75] Under U.S. sanctions regulations, foreign subsidiaries of American companies are able to trade or engage in business in Iran.

Petrochemicals

Iran is the second largest manufacturer of petrochemicals in the Middle East, following Saudi Arabia. About half of Iran's petrochemical product sales are for its domestic market.[76] In an attempt to diversify its exports, Iran also is building up its petrochemicals industry.[77] The industry reportedly faces some challenges from state intervention and price-fixing. Additionally, international sanctions have reduced commercial banks' willingness to finance international deals to build the petrochemical sector.[78]

Financial Sector

Iran's financial sector has been heavily dominated by large, public banks since the nationalization of the banking system after the 1979 revolution. Over the past couple of decades, Iran has engaged in some privatization and liberalization of its financial sector. In 2001, Iran's Central Bank approved licenses for three full functioning private banks. Efforts toward privatization have been thwarted frequently by the Guardian Council.

Iran's Central Bank, the Bank Markazi, technically is an independent institution. However, the Iranian government has direct control over lending and investment activities of commercial banks. The Bank Markazi, Iran's Central Bank, is not able to conduct a "proactive" monetary policy and has no control over the government's fiscal policy. In addition, the Central Bank is limited in its ability to issue direct instruments to combat inflationary pressures. The Central Bank must obtain approval from the *Majlis* in order to issue participation papers.[79]

State-owned banks are considered by many to be poorly functioning as financial intermediaries. Private banks are hampered by extensive regulations and the government's populist policies, including controls on rates of return and subsidized credit for specific regions. In May 2007, President Ahmadinejad capped lending rates to 12% for state-owned banks and 13% for commercial banks, despite strong opposition from the Central Bank. Setting interest rates below the rate of inflation reportedly has placed many commercial banks under financial duress.[80] In addition, most of the financial intermediaries' loan portfolios are comprised of low-return loans to state-owned enterprises and quasi-government agencies, such as the *bonyads*. Some believe that the financial system has stifled domestic business and has lowered Iran's attractiveness to foreign businesses.

Tehran Stock Exchange

In 1967, Iran began operating the Tehran Stock Exchange (TSE). With initially six companies, the TSE now lists over 300 companies. Capitalization through the TSE is permitted for the automotive, mining, petrochemical, and financial sectors. Since 2005, foreign investors have been able to participate in the TSE. Foreign investors are permitted to hold a maximum of 10% of shares of each company listed and are not allowed to withdraw their capital for three years after purchases. However, foreign activity in the TSE is low. Aside from concerns about the international tensions associated with Iran's nuclear standoff, low foreign activity may also reflect concerns about liquidity, transparency, and the poor legal environment protecting foreign holdings.[81]

The TSE index performed strongly between 2000 and 2004, but declined following President Ahmadinejad's election in 2005. During the 2007, the TSE market stabilized, but was still 20% lower than before Ahmadinejad came into power.[82] At the end of 2007, TSE market capitalization stood at $46 billion.[83] It is hoped by the Ahmadinejad government that privatization plans will help to revive the stock market.

Because the TSE is not extensively integrated into the world market, it has been somewhat insulated from the global financial crisis. However, between October and December of 2008, the TSE fell by nearly 20%, with investor confidence shaken by the global economic turndown and the impact of declining oil prices on Iran's economy.[84]

Financial Sanctions

The U.S. Department of the Treasury has employed targeted financial measures against Iran. The United States is attempting to isolate Iran from the international financial and commercial system in an effort to promote policy change in Iran regarding its nuclear program and purported terror financing. The United States also hopes that financial isolation will limit Iran's resources for its nuclear program and its alleged support for terrorist organizations. In congressional testimony, the Treasury Deputy Assistant Secretary for Terrorist Financing and Financial Crimes in 2008 stated, "Iran utilizes the international financial system as a vehicle to fund these terrorist organizations... the Iranian regime operates as the central banker of terrorism, spending hundreds of millions of dollars each year to fund terrorism."[85]

Several major Iranian banks are under U.S. and U.N. sanctions. Under E.O. 13224,[86] the Treasury has designated several Iranian entities for supporting terrorism. On October 25, 2007, the Treasury designated Bank Saderat, a major Iranian state-owned financial institution, for terrorism support. Iranian authorities contend that two external audits of Bank Saderat conducted in Lebanon and London found no evidence of such allegations.[87]

On January 9, 2007, the Treasury sanctioned Bank Sepah, another major Iranian financial enterprise, under E.O. 13382[88] for assisting with Iran's missile program.[89] U.N. Security Council Resolution 1747 named Bank Sepah and Bank Sepah International as financial institutions involved in financing nuclear or ballistic missile activities. Subsequently, on October 25, 2007, under E.O. 13382, the Treasury Department sanctioned Bank Melli and Bank Mellat, other major Iranian financial institutes, as WMD proliferators or supporters.[90] In June 2008, the European Union also decided to sanction Bank Melli.

In a signal to Arab countries, the United States sanctioned the Bahraini Future Bank B.S.C. in March 2008 under E.O. 13382 for reportedly assisting in Iran's nuclear and missile programs. The United States contends that Future Bank B.S.C. is controlled by the embargoed Bank Melli.[91]

In a move to further restrict Iran's access to the U.S. financial system, the Treasury revoked the "U-turn" license for U.S. financial institutions on November 6, 2008.[92] With respect to Iran, "U-turn" fund transfers are financial transactions that pass through the U.S. financial system only en route from one offshore non-Iranian financial institution for another, conducted for the direct or indirect benefit of the Iranian government, banks, or individuals. Previously, U.S. financial institutions were allowed to process such financial transactions.

The United States and some European countries assert that certain Iranian banks and their branches are attempting to circumvent international financial sanctions in order to engage in proliferation-related activity and terrorist financing. Iranian government officials have denied these claims.[93]

Financial sanctions reportedly have affected the profitability of Iranian banks and damaged Iran's credit ratings. Financial intermediaries have faced challenges financing development projects, such as building oil infrastructure.[94] Iran is taking steps to protect its foreign assets from future international sanctions. For instance, Iran reportedly has started shifting billions of dollars from European banks to Iranian and Asian banks and purchasing gold and equities. However, some economists express concern that Asian banks may not be reliable because of their close relationship to Europe's economy.[95]

Money Laundering

Iran's financial system may be vulnerable to money laundering. Since 2002, the Central Bank of Iran has engaged in efforts to combat money laundering. In January 2008, Iran passed its first anti-money laundering law, which criminalized money laundering. Critics contend that Iran's money laundering framework is not adequate to combat the terrorist financing through Iran's financial system. There is international concern that these vulnerabilities pose a threat to the international financial system.

On March 3, 2008, the U.S. Treasury's Financial Crime Enforcement Network (FinCEN) issued a statement emphasizing concern about ongoing deficiencies in Iran's efforts to combat money laundering and the financing of terrorism through its financial system. The U.S. Treasury advisory stated that, using state-owned banks, Iran "disguises its involvement in proliferation and terrorism activities through an array of deceptive practices specifically designed to evade detection." Of particular concern to the U.S. Treasury is that Iran's central bank and commercial banks have requested their names to be removed from international transactions in order to make it more difficult to track their involvement.

The Treasury advisory noted 59 major Iranian banks or their branches in international financial cities that pose threats, including Iran's central bank. None of the banks listed currently face U.N. or U.S. sanctions.[96] The advisory encouraged all financial institutions to consider the risks associated with doing with the specified Iranian financial institutions.[97] Additionally, the Financial Action Task Force (FATF), a Paris-based "international financial watchdog," called on its 34 member states to encourage banks to monitor their financial interactions with Iran. The FATF alleges that Iran has not taken adequate actions to combat money laundering and terror financing.[98] Iranian officials assert that the Bank Markazi complies with international best practices and that it vigilantly regulates domestic financial institutions.

Informal Finance Sector

Many Iranian businesses and individuals also rely on *hawala*, an informal trust-based money transfer system that exists in the Middle East and other Muslim countries. *Hawala* transactions are based on an honor system, with no promissory instruments exchanged between the parties and no records of the transactions. Some analysts consider the *hawala* system as particularly susceptible to terrorist financial transactions.

Since the imposition of recent U.S. and U.N. financial sanctions on Iran, the use of *hawala* by Iranians reportedly has increased. It is considered by many Iranians to be a more cost-effective way to transfer money in light of the added expenses incurred through working through the formal financial system in light of the sanctions. According to a Iranian merchant, "If we wanted to send money through the banking system it would cost a small fortune, so we move money to dealers and they send the money through Dubai to China." While some assert that the use of *hawala* shows that Iran is able to circumvent international sanctions successfully, others suggest that the increased use of *hawala* is a sign of the sanctions' effectiveness in making it more difficult for Iran to finance transactions.[99]

International Trade

International trade contributes significantly to Iran's economy and has increased dramatically over the past few years. Between 2004 to 2007, Iran's total trade in goods (exports plus imports) nearly doubled, reaching about $147 billion in 2007. Iran enjoys a growing and positive trade balance in goods, benefitting from high international oil prices. This trade surplus registered at about $36 billion in 2007. Exports totaled about $91 billion, while imports reached about $55 billion that same year (see Table 3). Some analysts point out that Iran's trade with the world may actually be higher due to transshipment or black market trade. Overall, Iran's external sector position has strengthened in recent years. The current account balance reached an estimated $29 million in 2007, about 10% of Iran's GDP at market price.[100] The current account surplus is expected to decline in 2009 with the drop in international oil prices.

Major Goods Traded

Oil and gas exports dominate Iran's export revenues, constituting about 80% of total exports and are the most important source of foreign exchange earnings for the country. Other major export commodities are petrochemicals, carpets, and fresh and dried fruits. Top destinations for Iran's non-oil exports, including natural gas liquids, are the United Arab Emirates (UAE), Iraq, China, Japan, and India. Major imports for Iran include gasoline and other refined petroleum products, industrial raw materials and intermediate goods used as manufacturing inputs, capital goods, food products, and other consumer goods.

Oil Exports

Despite low production levels, Iran remains the fourth largest exporter of crude oil worldwide, after Saudi Arabia, Russia, and Norway.[101] Net crude and product exports in 2006 totaled 2.5 million barrels per day and $54 billion revenues. Top export markets for Iran are Japan, China, India, South Korea, and Italy. More than 40% of the world's oil traded goes through the Strait of Hormuz, a channel along Iran's border. The Strait of Hormuz is considered a global "chokepoint" because of its importance to global energy security. It is a narrow channel with a width of only 21 miles at its widest point through which large volumes of oil are shipped.[102]

Table 3. Iran Merchandise Trade, FY2004-FY2007 (millions of U.S. dollars)

Merchandise	2004	2005	2006	2007[a]
Exports	44,364	64,366	75,537	91,289
Oil and gas	36,827	53,820	62,458	76,858
Non-oil and gas	7,537	10,546	13,079	14,431
Imports	38,199	43,058	49,292	55,352
Gasoline[b]	2,639	4,190	5,745	6,135
Trade Balance	6,165	21,281	26,245	35,938
Total Trade	82,563	107,451	124,829	146,641

Sources: IMF, "Islamic Republic of Iran: 2008 Article IV Consultation," August 2008, IMF Country Report No. 08/284. IMF, "Islamic Republic of Iran: 2006Article IV Consultation," March 2007, IMF Country Report No. 07/100.

a. All data for 2007 are estimated.

b. Gasoline imports data, based on the "2006 Article IV Consultation," are preliminary for 2005 and projected for 2006 and 2007. The "2008 Article IV Consultation" does provide an update on gasoline imports data.

Oil revenue accounts for the majority of export earnings and represents the bulk of government revenue (about 40%). Iran's dependence on oil export revenues makes the country highly susceptible to the volatility of international oil prices. The quadrupling of global oil prices since 2002 has given Iran enormous economic and political leverage, cushioning the extent to which Iran's economy has been affected by international sanctions and domestic policy mismanagement. However, the recent decline in oil prices may highlight weaknesses in Iran's economy.

There is debate about the extent to which the recent decline in oil prices will affect Iran's economy. Oil price declines would reduce government revenue and spending and potentially increase Iran's vulnerability to sanctions. Oil price drops also would affect the private sector, as Iran imports a significant portion of its capital and machinery goods from abroad. A fall in oil prices and subsequent economic downturn may increase political dissent among Iranians, already facing high unemployment and inflation levels.

Refined Petroleum Imports

Despite Iran's vast oil reserves, the country must import close to half of all refined petroleum products to meet domestic consumption needs. Iranian gasoline imports were projected to total $5.7 billion in FY2006 and $6.1 billion in FY2007.[103] Extensive government subsidies on gasoline have contributed to high gasoline consumption rates. Many analysts contend that high subsidies do not give Iranians an incentive to conserve. In addition, there has been an increase in vehicle sales, particularly of fuel-inefficient older models. Import levels are also high because Iran has limited domestic refinery capacity to produce light fuels.[104] However, gasoline's share of imports has fallen recently, from 18% in FY2005 to 6% during the first eleven months of FY2007, according to Iran's Customs Administration. In June 2007, the government implemented a gasoline rationing system to reduce gasoline consumption. This policy was extremely unpopular and led to public riots, but has led to a drop in gasoline consumption. Oil consumption also is declining as consumers are moving more toward natural gas use.

Major gasoline suppliers to Iran historically have been India, Turkmenistan, Azerbaijan, the Netherlands, France, Singapore, and the United Arab Emirates. Iran also imports gasoline from multinational companies (MNCs), particularly Europe-based wholesalers. Based on data from 2005 through 2006, Turkmenistan was Iran's only supplier of natural gas.[105] In 2006, Vitol, a MNC based in Switzerland, supplied Iran with 60% of its total gasoline cargo imports. In December 2007, Vitol reportedly declined to renew long-term contracts with Iran, but still provides gasoline to Iran on the spot market. Major gasoline suppliers to Iran include BP, Royal Dutch/Shell (Netherlands), Total (France), Lukoil (Russia), and Sinopec (China). In January 2009, India's Reliance Industries Limited (RIL) reportedly agreed to terminate gasoline sales to Iran once its current contractual obligations expire.[106] Previously, some Members of Congress called on the U.S. Export-Import Bank to rescind two loan guarantees worth $900 million authorized to RIL, in support of the RIL's petroleum refinery equipment and services ($500 million) and for gas development and exploration in India's Bay of Bengal region ($400 million).[107]

In addition, Venezuela supplies small quantities of gasoline from time to time in a show of political solidarity with Iran. Iran and Venezuela have sought to counter U.S. global influence and strengthen their own international standing and reputation through strategic alliances.[108]

Table 4. Major Export Markets and Sources of Imports for Iran, FY2007 (millions of U.S. dollars)

Country	Total Trade	Exports	Imports	Trade Balance
China	20,135	12,188	8,017	4,101
Japan	13,064	11,599	1,465	10,134
South Korea	8,421	5,139	3,282	1,857
Italy	8,039	5,215	2,824	2,391
Turkey	7,539	6,013	1,526	4,487
Germany	6,066	621	5,445	-4,824
UAE	5,915	747	5,168	-4,421
France	5,334	3,069	2,265	804
Russia	3,272	282	2,990	-2,708

Source: IMF, Direction of Trade Statistics

Key Trading Partners

In 2007, Iran's top overall trading partner was China. Iran's next overall largest trading partner is Japan, followed by Italy, South Korea, and Germany. Significant export markets for Iran are China, Japan, Turkey, and Italy. Major merchandise suppliers for Iran include Germany, China, the United Arab Emirates, and South Korea (see Table 4).

Iran's trading relations have changed over time as international concern over Iran's nuclear program has affected economic activity. Historically, Iran has had strong trade ties with Germany. In 2007, Iran's exports to Germany increased by 50%, while imports from Germany declined by 4%. Despite the increase in exports, Iran's total trade with Germany dropped by 0.3% in 2007.[109] Along with Germany, other European countries also have scaled down trade with Iran lately.

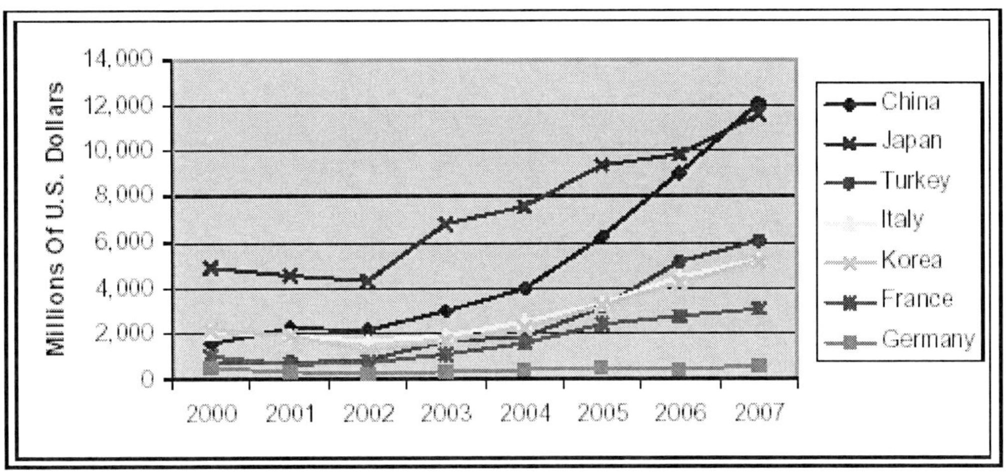

Source: IMF, Direction of Trade Statistics

Figure 1. Iran's Exports to Select Countries, FY2000-FY2007

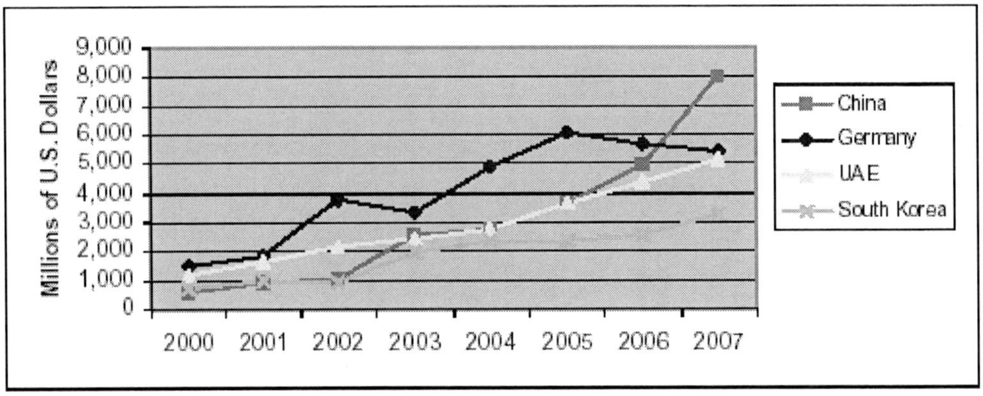

Source: IMF, Direction of Trade Statistics

Figure 2. Iran's Imports From Select Countries, FY2000-FY2007

Iran's trade has shifted from Western countries to the developing world. Asian countries, particularly China and Japan; Russia and other Central Asian countries; and Middle Eastern countries, especially the United Arab Emirates, have emerged as growing and important trading partners for Iran. Figure 1 shows the shift in Iran's exports to trading partners is shown in from FY2000 to FY2007, while Figure 2 shows the change in Iran's import relationships during the same period.

Facing challenges in trading with Western countries, Iran has sought to strengthen ties with Asian countries, most notably China and Japan. Between 2000 and 2007, total trade between Iran and China grew more than nine-fold, reaching about $20 million in 2007. China was Iran's biggest exporter in 2007, followed by Japan and Turkey. Iran benefits from low-cost imports from China. Major Chinese exports to Iran include mechanical and electrical equipment and arms. Iran's growing trade relationship with China also may be rooted in strategic reasons, such as China's position as one of five permanent UNSC members.[110]

Russia is becoming an important trading partner for Iran. While Iran-Russia trade is low compared to Iran's trade with other countries, this relationship has grown significantly. Iranian imports from Russia more than tripled from 2000 to 2007 and registered at about $3 million in 2007. In addition, several countries in central Asia have declared their interest in boosting economic engagement with Iran, including Tajikistan.[111] Iran, Turkey, and Azerbaijan have held discussions on building a joint railway through the three countries in order to enhance relations, trade, and travel.[112]

Iran also has pursued increased integration with its neighbors in the Middle East. Merchandise trade with the Middle East has grown. Arab nations may be weary of Iran's nuclear ambitions, but they appear to value trade and investment relations with Iran. Many are hoping that positive economic engagement with Iran will mitigate international tensions over Iran's nuclear ambitions.

United Arab Emirates and Transshipment Trade

The UAE is a major trading partner for Iran, with trade largely dominated by UAE exports to Iran. According to the UAE Ministry of Economy, total non-oil trade between UAE and Iran was upwards of $6 billion in 2006. The bulk of merchandise supplied to Iran by the UAE is believed to be products imported into the UAE from foreign markets, including the United States, European Union, China, and India, and subsequently repackaged for shipment to Iran. Iran represents about 14% of the UAE's total exports, including re-exports. According to the Dubai Chamber of Commerce and Industry, re-exports accounted for about 60% of the UAE's $6.8 billion trade with Iran in 2006. The rest of the trade came from the UAE's free trade zones. Iran's foreign investments in Dubai are more difficult to verify, but may have neared $300 billion, about a quarter of Iran's total foreign investments.[113]

The UAE thrives as a central re-exporting and distribution center in the Persian Gulf because of its low tax rates, free trade zones, lower delivery times, enhanced handling and service capacity, and a perception of lax export controls. Dubai, in particular, is Iran's economic lifeline to the rest of the world. Through Dubai, Iran is able to import goods that the country cannot import directly due to international and U.S. sanctions. Although U.S. businesses are outlawed from operating in Iran, many reportedly can circumvent U.S. sanctions by sending their investments through Dubai.

The United States has called on the UAE to make its export controls more stringent. Generally speaking, the UAE has resisted such pressure. In recent months, the UAE appears to be taking actions to regulate trade and investment relations with Iran in a more stringent manner. In September 2007, the UAE passed a law permitting it to place restrictions on dual-use technologies, chemical and biological weaponry, and military equipment. The UAE used the new law for the first time to impound a vessel at Jebel Ali that was delivering merchandise to be transshipped to Iran.[114] About 40 Iranian companies were closed in 2007 based on UAE efforts to reduce trade in goods with potential "dual use."[115]

Financial institutions in the UAE reportedly are restricting Iran's access to credit, making it harder for Iranian businesses to trade goods with the UAE. Some UAE banks reportedly have frozen the assets of Iranian firms and have reduced opening letters of credits to Iranian businesses.[116] UAE-based banks may be wary of reputational and financial risks associated with such transactions. Consequently, some Iranian businesses have had to shift to other regional banks or have had to engage in cash-based transactions, raising the costs of goods on the end-user.[117] Dubai continues to be an important transshipment hub for Iran, but some parts of the Iranian business community are concerned about the potential implications of a more stringent UAE approach to commercial ties with Iran. There is a possibility that trade diversion to Iran may take place through other countries if the UAE is perceived as a hostile business environment.[118]

U.S.-Iranian Trade

U.S. trade with Iran is limited, receding drastically with the 1987 U.S. ban on imports from Iran and the 1995 ban on U.S. exports to and investments in Iran. Before 1995, major U.S. exports to Iran included machinery and industrial equipment. U.S. exports virtually came to a standstill with the 1995 embargo on U.S. trade and new investment in Iran. Sanctions were relaxed to a certain extent in 2000, with the election of President Khatami in Iran. While U.S. trade with Iran is low compared to U.S. trade with other countries, there has been notable growth in U.S.-Iranian trade in recent years (see **Table 5**).

U.S. exports to Iran include pharmaceuticals, tobacco products, wood pulp, optical and medical instruments, and wheat. The top U.S. imports from Iran are textile and floor coverings, art and antiques, fish and seafood (caviar), prepared meat and fish, and edible nuts and foods (pistachios). There is evidence that Iran is able to obtain embargoed U.S. goods through the re-export trade, mainly through Dubai.[119]

U.S. sanctions against Iran may curtail U.S. economic activity, imposing costs on American workers and businesses and reducing U.S. exports.[120] U.S. businesses have expressed concerns about U.S. measures against companies that are unable to control re-exports of high-technology goods to Iran and other targeted countries. Others have noted that U.S. policies in Iran may deprive the United States of significant business opportunities in Iran. Europe, China, India, and Russia are stepping in and taking advantage of Iran's sizeable market and untapped potential. Proponents of sanctions contend that the security, reputational, and financial risks associated with doing business with Iran outweigh the economic benefits.

In general, entities targeted by U.S. sanctions do little business with the United States. Consequently, the United States must depend on other countries to reduce trade and investment with Iran in an effort to change Iran's policies. Such sanctions would have little effect on U.S.-Iran trade since such trade is already limited. However, the action would send a strong signal to foreign countries and may hurt Iran's trade with major trading partners.[121]

International Sanctions and International Trade

Since 2006, European Union countries, including France, Germany, and Britain, have curtailed export credits to companies doing business in Iran.[122] For example, in 2007, German export credits backing trade with Iran totaled about $730 million, about half the value of German export credits in 2006 and one-fifth that in 2004.[123] Germany does not actively dissuade companies from doing business in Iran, but it is conducting extra scrutiny of export authorizations requests and evaluating the financial risks of doing business with Iran more closely.[124] Despite the scaling back of German official export credits for trade with Iran, German exports to Iran reportedly grew by 10% in 2008, reaching $55 billion.

Some large European banks have reduced businesses with sanctioned Iranian bodies. For instance, Germany's Commerzbank and Deutsche Bank, have been reducing or stopping business with Iran. The United Kingdom's HSBC and Standard Chartered have also reduced business with Iran.[125] Many European banks that have curtailed business with Iran are leaving offices open on a minimal basis in case there is a change in the international climate towards Iran.[126]

The merchant class has particularly been hurt by the international sanctions. Iranian businessmen reportedly have increased difficulty opening bank accounts abroad and getting foreign banks to honor letters of credit. According to Iranian officials, over half of the banks in Dubai no longer provide credit to businesses based in Iran.[127] As Iranian businesses experience setbacks in obtaining trade financing from international banking partners, they may turn to lesser known banks or to other banking partners not susceptible to international pressure, but potentially raising the cost of business. In particular, the Islamic Republic has turned toward banks in Gulf Cooperation Council (GCC) countries.

Table 5. U.S.-Iranian Trade, FY2000-FY2008 (Millions of U.S. Dollars)

Year	U.S. Exports	U.S. Imports	Balance
2000	17	169	-152
2001	7	143	-136
2002	32	156	-124
2003	99	161	-62
2004	85	152	-67
2005	96	175	-79
2006	86	157	-79
2007	145	173	-28
2008	683	104	581

Source: U.S. Census Bureau, Foreign Trade Statistics

Sanctions against Iran's shipping industry also may pose challenges for Iran's trade with the rest of the world. U.S. sanctions imposed on the shipping company Islamic Republic of Iran Shipping Lines (IRISL) and its affiliates likely will have limited effect on Iranian trade, as the IRISL and its subsidiaries do little business in U.S. waters, the bulk being in China and other Asian countries. U.N. sanctions, which allow for the inspection of Iranian vessels suspected of carrying embargoed goods, may have more impact on Iran's international trade. To avoid being stopped, Iranian vessels may choose to sell under foreign flags, which are subjected to a import duty by the Iranian government. While the Transportation Minister exempted Iranian vessels sailing under foreign flags from the import tax, the provision has yet to be implemented. Consequently, if the U.N. sanctions are followed closely, they may ramp up the cost of business for Iranian shipping companies and raise the price of imports for Iranian consumers.[128]

Trade Liberalization

In 1995, Iran became a WTO observer state and, since then, has repeatedly put forth applications to become a permanent WTO member. Accession to the WTO is a stated priority of the Iranian government. Iran cites the more favorable treatment that WTO members give to one another and competition from Asian countries in textiles and manufactures as important challenges to Iranian exports.[129] The United States repeatedly blocked Iran's bids to join the WTO over concerns about Iran's nuclear program and support for terrorist activities. On the other hand, many European Union countries and developing countries have supported Iran's accession. Iran and many other countries maintain that WTO membership should not be based on political reasons, but rather, on economic and business grounds.[130]

In a significant policy shift toward Iran in May 2005, the United States agreed to stop blocking Iran's attempts to join the WTO as part of economic incentives to Iran to resolve the nuclear program issue. However, the most recent negotiations for accession have ceased because of political reasons and Iran continues to not be a member of the WTO.[131] The WTO accession process is lengthy and some Iranians have expressed concern that domestic momentum for the reforms necessary for accession has waned. Iran, along with Russia, now remain the two largest economies outside of the WTO.

INTERNATIONAL FINANCIAL FLOWS

Foreign Exchange Reserves

Iran's foreign exchange reserves, which include the Oil Stabilization Fund (OSF), tend to follow international oil prices. Based on IMF estimates, Iran's international reserves grew from $60.5 billion in FY2006 (10.2 months of imports cover) to $81.7 billion in FY2007 (11.5 months of imports).[132] For FY2008, Iran's international reserves totaled $100 billion (14 months of imports).[133] Owing to the recent drop in oil prices, Iran's international reserves may shrink with declines in oil income. There is concern that domestic economic mismanagement has reduced funds available through the OSF to smooth economic vulnerabilities facing Iran during the current global economic and financial crisis.

The composition of Iran's foreign reserves has changed lately. In December 2007, Iran stopped accepting payments in U.S. dollars for oil export purchases by foreign countries and is shifting to other currencies, such as the euro and the yen. Iran also called upon other OPEC members to shift away from the dollar in favor of other currencies during a November 2007 OPEC summit. Aside from Venezuela, all other member states opposed the switch.[134] The Central Bank also is reducing the proportion of dollars in its foreign reserves and diversifying to other currencies.[135]

Iran claims that this move away from dollars is in response to the recent slide of the U.S. dollar, and cited the dollar as an unreliable currency. Others contend that the shift away from the dollar is a retaliatory measure against the United States. Iran's currency diversification also may be an effort to seek independence from the dollar in light of punitive U.S. measures against Iran, which have contributed to Iran's isolation from the international financial system. This may mitigate U.S. ability to pressure other countries to follow along with sanctions against Iran.[136] For the United States, these recent events may raise questions about the long-term viability of the U.S. dollar as the global oil currency and have potential implications for the U.S. economy. U.S. massive current account deficits are financed by foreign countries' purchase of U.S. dollar denominated assets.

Foreign Investment in Iran's Economy

As the most populous country in the Middle East and with vast natural resources, Iran potentially has a significant market for foreign businesses. However, foreign direct investment (FDI) in Iran historically has been low. In 2007, the flow of FDI into Iran stood at $754 million, compared to FDI inflows of $24.3 billion to Saudi Arabia, $22.0 billion to Turkey, and $11.6 billion to Egypt. While low, FDI inflows to Iran have fluctuated in recent years. Iran's FDI in Iran was more than double 2006 inflows of $317 million, but down from 2005 inflows of $917 million.[137]

A stringent domestic regulatory environment and government reluctance to allow foreign investment have contributed to low levels of FDI. For instance, in the energy sector, foreign investment is believed to be limited due to Iran's buy-back system. Under this system, international oil companies that contract with an Iranian affiliate pay a fee—such as an "entitlement to oil or gas from development operation." In 2006, buybacks were projected to reach $500 million.[138] Despite some widespread resistance within the Iranian parliament (the *Majlis*) and other parts of government, Iran is slowly letting investors into the country.

However, international sanctions and political uncertainty have clouded Iran's economy and have made foreign business and investors wary about economic involvement in Iran. U.N. and some U.S. sanctions are targeted toward obstructing Iran's development of its oil and gas sectors in order to constrain Iran's resources for uranium enrichment and alleged terrorist financing.[139] Foreign companies have had difficulty obtaining financing due to U.S. Treasury Department pressure on international banks to cut off ties with Iran.[140] Foreign companies also have limited investment in order to avoid U.S. opposition and to maintain good relations with the United States. International investors reportedly have withdrawn from development projects in the country, such as in the oil and gas, shipping, and automotive industries.[141] Iran faces a problem of significant domestic capital flight abroad, particularly to the UAE.

International energy companies that have decided to suspend development projects in Iran include British Petroleum, Total, Royal Dutch Shell, Repsol YPF, StatoilHydro, and Eni. Some companies have decided to continue current projects, but to not engage in any future projects with Iran for the time being.[142] As some European companies have scaled down energy sector development projects, other European partners are stepping in (see Table 6 for selected recent deals negotiated by Iran). For instance, Iran signed gas deals in 2007 with Switzerland and Austria, deals that were strongly opposed by the United States. The State Department is evaluating the deals for possible violations of the Iran Sanctions Act.[143] In addition, Germany recently announced plans to build three LNG plans in Iran.

With Western involvement in Iran's energy sector tenuous, Iran has been turning toward Asian countries, such as China and Pakistan; Russia and Central Asian countries; and regional partners, such as Bahrain and Turkey. While new agreements have been negotiated, their successful completion has been slow. Many speculate that the deals are not finalized because of international concerns over Iran's nuclear enrichment program and the specter of sanctions.[144] According to a GAO report, State and Treasury officials assert that U.S. sanctions have contributed to a delay in foreign investment in Iran's hydrocarbon sector.[145] The Iranian government contends that sanctions and international pressure have not slowed down foreign investment in Iran's gas sector.

Others point out that LNG contracts with Asian and Eastern European countries may not be able to deliver the same quality as Western contracts. For instance, despite the possible termination of Shells' LNG project with Iran, Iran appears to be "keep[ing] open the option of enlisting Shell's technical and marketing know-how and financial input for an LNG project linked to a future phase of South Pars."[146]

Iran also is engaging in efforts to privatize nearly 50 state-run oil and gas companies, estimated to be worth $90 billion, by 2014 through the Tehran Stock Exchange. Both domestic and foreign investors would be able to buy shares. Privatization of these energy companies may make it easier for investors to circumvent U.S. sanctions, which complicate investors' ability to engage in business transactions with Iran directly.[147]

In the United States, there has been a growing grassroots movement to divest from Iran. Divestment is a decision to not hold stock in a company or bank that does business with Iran.

There is a call to remove current stock from such companies. States such as Louisiana and California recently have passed measures to divest from Iran.[148]

Table 6. Selected International Energy Deals Negotiated by Iran

Country	Notes
Switzerland	Swiss energy company EGL signed a 25-year LNG export deal with the National Iranian Gas Export Company (NIGC) in March 2007, reportedly valued at €18 billion. Switzerland will buy 5.5 billion cubic meters of Iranian natural gas each year, beginning in 2011. This would be Europe's second largest gas deal.[149] There is some skepticism that Iran will not be able to supply gas to Switzerland for the foreseeable future because no pipeline connects Iran to Europe at present. The State Department is evaluating the deal for possible violations of the Iran Sanctions Act.
Austria	Austrian partially state-owned energy company OMV signed letters of intent with Iran in April 2007 for Iran to supply Europe with gas, transactions reportedly valued at $23 billion. The State Department is evaluating the deal for possible violations of the Iran Sanctions Act.
Russia	In February 2008, Russian state gas company Gazprom announced a deal to establish a joint venture company to develop the offshore Iranian South Pars gas field. Iran would benefit from a build-up of its gas export infrastructure.[150]
China	A China National Offshore Oil Corporation (CNOOC) investment deal, reportedly valued at $16 billion, to develop Iran's North Pars gas field and to build a liquid natural gas (LNG) plant, was supposed to be signed on February 27, 2008 but has been delayed. Some analysts believe that China has been hesitant to finalize the deal because of international reaction to Iran's nuclear program and the tightening of United Nations sanctions. The state-operated National Iranian Oil Company (NIOC) and CNOOC signed a memorandum of understanding in December 2006 for the project, under which CNOOC would purchase 10 million metric tons per year of LNG for 25 years.[151] The United States has criticized China's pursuit of the deal with Iran. China has also looked into alternate suppliers, such as Qatar and Australia.[152]
Pakistan	NIGC is expected to finalize a natural gas export deal with Pakistan in April 2008, reportedly valued at $7.4 billion. Exports are set to begin in 2011. The gas would be transported through a "Peace Pipeline." The plan initially also included exporting gas to India, but negotiations have stalled over pricing. The United States has strongly opposed the pipeline and pressured India and Pakistan to halt the project.
Turkey	Iran is discussing gas production and export deals with Turkey. Under the plans, Turkey would assist in developing Iran's South Pars field in exchange for cash or natural gas. Gas would be shipped from Iran to Turkey and then to other parts of the world, including Europe, via a new pipeline that Turkey plans to build.[153]
Germany	In November 2008, the German company Steiner recently announced plans to build three LNG plants in Iran, reportedly valued at $147 million.[154] While there has been some international criticism of this decision, German authorities point out that the deal did not violate export controls of sensitive goods.[155] Iranian officials assert that it will continue to develop Iran's gas fields, even with foreigners pulling out investment.

Country	Notes
Malaysia	In December 2008, Iran signed gas deals reportedly valued at $14 billion with Malaysia's SKS Group. The deal included a contract to build an LNG plant.[156]
Bahrain	In December 2008, Iran and Bahrain drafted an agreement on the delivery of 1 billion cubic feet per day of natural to Bahrain.[157]
Russia	In March 2009, Gazprom, a Russian conglomerate, and the National Iranian Oil Company (NIOC) reportedly concluded a hydrocarbon swap agreement under which Gazprom agreed to deliver Turkmen gas to Iran in exchange for gas and petroleum products from Iran.[158]

International Loans and Assistance

World Bank

Iran receives loans from the World Bank. As of November 30, 2008, the net principle amount of World Bank loans totaled Iran $3.4 billion, of which $2.6 billion had been disbursed.[159] The World Bank currently has nine active portfolios in Iran, focused on reconstruction efforts. The World Bank's activity in Iran restarted in 2000, following a seven year halt. World Bank loans to Iran come only from the International Bank for Reconstruction and Development (IBRD), the Bank's market-rate lending facility. Iran is unable to borrow from the Bank's International Development Agency (IDA), a concessional lending and grant-making fund, because of its per capita GDP. The United States has not made any contributions to the IBRD, which lends to Iran, since 1996. Some lawmakers call for reducing U.S. contributions to the IDA in protest of IBRD lending to Iran. However, some question the merits of penalizing other countries that receive loans from the IDA.

In addition, the World Bank's International Finance Corporation (IFC) recently invested in Iran, providing a $5 million joint venture among a Iranian private bank, a French bank, and the IFC. In addition, Iran has joined the World Bank's Multilateral Investment Guarantee Agency (MIGA), which offers political risk insurance to foreign and domestic investors in Iran.[160]

Bilateral Official Development Assistance

In terms of bilateral official development assistance (ODA), major donor countries to Iran are Germany, France, the Netherlands, Norway, and Japan. On the whole, the United States does not provide foreign assistance, but does provide some humanitarian assistance, to Iran. For instance, USAID has provided disaster relief assistance following the earthquake that struck near the Iranian city of Bam on December 26, 2003. Total ODA given by countries of the Organization for Economic Cooperation and Development (OECD) to Iran amounted to $71 million in 2006 (see Table 7).

Table 7. Net ODA to Iran from OECD DAC Members, 2001-2006 (millions of U.S. dollars)

Donor	2001	2002	2003	2004	2005	2006
Austria	3.5	3.4	5.7	6.4	4.4	3.4
Germany	32.6	31.8	38.8	41.2	40.6	38.4
France	6.8	7.9	9.5	15.7	14.8	15.4
Japan	34.4	17.5	11.3	19.8	-2.5[a]	-7.3[a]
Netherlands	3.8	3.8	7.7	11.1	6.8	1.0
Norway	3.7	5.3	9.7	11.5	4.3	2.6
United States	—	0.2	0.5	4.8	3.8	2.3
Total DAC Countries[b]	90.8	81.5	102.1	138.9	78.2	70.8

Source: OECD, "Geographical Distribution of Financial Flows to Developing Countries," 2007 and 2008 editions.

a. Negative grants may be due to the return to the owner of unspent balances that were previously disbursed as grants.

b. OECD DAC members for which data is reported for are Australia, Belgium, Canada, Denmark, Finland, France, Germany, Greece, Ireland, Italy, Japan, Luxembourg, the Netherlands, New Zealand, Norway, Portugal, Spain, Sweden, Switzerland, the United Kingdom, and the United States.

Congressional Issues and Options

Members of Congress appear to be divided about the United States' course of action with respect to Iran. Some contend that the United States should pursue harsher measures against Iran, given the gravity of the real and potential threats posed by Iran's uranium enrichment program and terrorism financing. Others suggest that perhaps the United States should consider more positive engagement with Iran through rebuilding diplomatic ties and pursuing economic engagement with Iran, such as through Iran's accession to the World Trade Organization. They suggest that the Iranian state would be receptive to sincere positive engagement on the part of the United States.

Unilateral and Multilateral Approaches to Sanctions

There is debate about whether or not the United States should pursue more sanctions against Iran unilaterally or through U.N. action. Some lawmakers assert that U.S. unilateral efforts to pressure Iran may detract from building multilateral consensus to widen punitive measures against Iran through the United Nations. Some maintain that unilateral efforts also might reduce Iran's willingness to cooperate with the United Nations. Additionally, foreign countries, such as the Persian Gulf states, may not be as responsive to unilateral action by the United States as they would be to multilateral action, given the growing trade ties between the Gulf states and Iran. For instance, Arab diplomats note that while they would respect U.N. sanctions against business with Iran, it is difficult to comply with unilateral U.S. sanctions in light of growing trade relations with Iran.[161]

Others note that pursuing multilateral action can be a lengthy process and that it is difficult to find consensus among foreign countries with various competing interests, such as promoting international security and promoting economic growth through trade with Iran. The United Nations successfully passed the third round of sanctions against Iran only after watering them down to satisfy Chinese and Indian concerns. As industrializing countries with increasing energy demands and insufficient supplies, China and India view Iran as a critical energy supplier for their needs. Such short-term national interest priorities may override international long-term security concerns about Iranian alleged terrorist financing or nuclear technology development. Still, some lawmakers consider the recently-passed third U.N. resolution a good first step and support pushing for more punitive action through the UNSC. Others suggest that to the extent to which China and India engage in economic transactions with Iran may be muted somewhat by the two countries' ties with the United States.

Impact of Sanctions on Iran's Economy and Policy

Analysts debate the impact of sanctions on Iran's economy. International tensions associated with Iran's nuclear program and alleged financing for terrorist organizations undoubtedly have complicated Iran's business environment. Some analysts point to Iran's low levels of foreign investment, difficulties obtaining trade finance, and challenges in developing its oil and gas sectors as evidence of the impact of sanctions.

On the other hand, according to a GAO report, "Iran's global trade ties and leading role in energy production make it difficult for the United States to isolate Iran and pressure it to reduce proliferation activities and support for terrorism." The Peterson Institute for International Economics (IIE) writes that sanctions increasingly have been unsuccessful as globalization has allowed embargoed countries to find other suppliers and export destinations for trade and investment.[162] Sanctions may not raise the costs to the point that they are crippling to the Iranian's trade and financial interactions with the rest of the world. Iran reportedly is able to circumvent the trade ban by transshipment of U.S. exports through other countries, such as the UAE. In addition to transshipment, analysts also note that international sanctions may simply divert Iran's trade to other countries that do not enforce sanctions against Iran.

In addition to the impact of sanctions on Iran's economy, some lawmakers question the effectiveness of sanctions, noting that despite thirty years of sanctions, the United States has not been able to significantly shift the Iranian government's policies. Previous studies have found that sanctions have little impact on government policy and that, rather, they tend to hurt the population of a country. In congressional testimony, one observer stated, "In a broader sense, sanctions often end up hurting ordinary people while having little impact on the government leaders we are trying to influence."[163] The recent enforcement of targeted financial measures appears to signal an effort to avoid the drawbacks of past broader trade sanctions efforts and to concentrate pressure on key negative actors. Despite the narrow focus of targeted financial sanctions, the effects of these sanctions may spill over to the broader Iranian populace. There is uncertainty about how sanctions affect the elite, and how elite views may spillover into government policy.[164]

Congress may choose to follow with GAO's assessment and require the U.S. Treasury and State to collect data to assess the economic impact of sanctions on Iran. According to a recent GAO report, U.S. economic sanctions on Iran have had affected Iran, but the extent of these effects on Iran's economy and behavior are difficult to gauge. The GAO notes that assessment of the impact of sanctions is challenging because of a lack of data collection by the U.S. government and baseline information for comparability.[165]

Action in the 110th Congress

In the 110th Congress, several bills were passed in the House related to Iran. House-passed bills encouraged tighter sanctions against Iran, but noted that such action does not indicate congressional support for U.S. military action against Iran. The following were some of the major pieces of legislation proposed by lawmakers:

- H.R. 957, "To amend the Iran Sanctions Act of 1996 to expand and clarify the entities against which sanctions would be imposed," would have stiffened existing sanctions against Iran. The bill was passed by the House on July 31, 2007 and referred to Senate committee on August 3, 2007.
- H.R. 2347, "Iran Sanctions Enabling Act of 2007," and the corresponding Senate version of the bill (S. 1430) would have encouraged divestment from companies that conduct business with Iran. The Bush Administration opposed H.R. 2347 on the grounds that it may interfere with U.S. foreign policy efforts. The bill would have allowed for sanctions against countries such as China, Russia, and France for conducting business with Iran.[166] H.R. 2347 was passed by the House on July 31, 2007 and referred to Senate committee on August 3, 2007.
- H.R. 1400, "The Iran Counter-Proliferation Act of 2007," and its companion bill, S. 970, would have expanded economic sanctions against Iran and remove the presidential waiver in the Iran-Libya Sanctions Act. H.R. 1400 was passed by the House on September 25, 2007 and referred to Senate committee on September 26, 2007.
- H.R. 2798 was a more narrowly targeted measure against Iran. It would have prohibited any assistance by the Overseas Private Investment Corporation (OPIC) to individuals who have finance or investment ties to countries that are state sponsors of terror. The bill would have targeted Iran, North Korea, and Sudan. The bill was passed by the House on July 23, 2007 and was ordered to be reported out to the Senate Committee on Foreign Relations and placed on the Senate Legislative Calendar on March 4, 2008.
- H.R. 1357, "To require divestiture of current investments in Iran, to prohibit future investments in Iran, and to require disclosure to investors of information relating to such investments," was referred to House subcommittee on June 5, 2007.
- H.R. 7112, "Comprehensive Iran Sanctions, Accountability, and Divestment Act of 2008," would have widened current sanctions to U.S. firms with foreign subsidiaries doing business in Iran; would have encouraged businesses to divest from Iran; and would have imposed penalties on countries that are involved in transhipment, re-

exportation, or diversion of sensitive goods to Iran. H.R. 7112 was passed by the House on September 26, 2008 and referred to House committees. The related Senate measure, S. 3445, was introduced on September 26, 2008.

Action in the 111th Congress

In the 111th Congress, several bills have been introduced related to Iran in a move to expand economic and diplomatic pressure on Iran. H.R. 1327, the "Iran Sanctions Enabling Act of 2009" (Barney) would support recent actions taken by some state and local governments and education institutions to divest their pension funds from persons and companies that invest more than $20 million in Iran's energy sector. The bill would prohibit legal action against asset managers who divest and would require states to take several measures when considering divestment, including providing the targeted entity with an opportunity to demonstrate that it does not have investment activities in Iran meeting the $20 million threshold. Supporters of the legislation assert that a divestment policy provides the United States with significant political and financial leverage against Iran. Critics argue that a divestment strategy would hamper U.S. diplomatic negotiations with Iran. The United States also may risk trade retaliation from countries whose companies are facing divestment. Others also argue that U.S. divestment efforts may not significantly affect the financial operations of foreign companies to the extent that might prompt them to reconsider their business activities in Iran. H.R. 1327 was introduced on March 5, 2009, and a committee hearing was held on March 12, 2009. On April 28, 2009, H.R. 1327 was ordered to be reported by voice vote. In the 110th, the House passed two similar proposals.

Some lawmakers have advocated targeting Iran's dependency on imports of refined petroleum products. S. 908, the "Iran Refined Petroleum Sanctions Act," would permit sanctions to be imposed on investments of $20 million or more that directly and significantly contribute to the enhancement of Iran's ability to develop Iranian petroleum resources. In addition, the bill would permit sanctions on the exportation of refined petroleum resources to Iran and activities that support the exportation of refined petroleum resources, including shipping, insurance, and financing activities. S. 908 was introduced on April 28, 2009, and referred to committee. Its companion bill, H.R. 2194, was introduced on April 30, 2009, and referred to committee. Supporters of such punitive options assert that they will place pressure on the Iranian government, given Iran's lack of refining capacity and dependence on gasoline imports. Others express concern that such action would adversely affect global energy supplies and ramp up prices for U.S. consumers.[167] Some critics also maintain that such an action would target the Iranian populace more than the regime.

End Notes

[1] Economic data for this chapter are drawn from data from the International Monetary Fund (IMF). As a member of the IMF, Iran reports on its economy to the IMF. The economic data are limited in their means of independent verification by the IMF. In addition, this chapter relies on data from the Economist Intelligence Unit (EIU) and

Global Insights, international economic research and forecasting agencies. U.S. government sources of data include the Central Intelligence Agency for general economic indicators and the Census Bureau for trade data.

[2] Abdelali Jbili, Vitali Kramarenko, and José Bailén, "Islamic Republic of Iran: Managing the Transition to a Market Economy," IMF, 2007, pp.1-5.

[3]Ibid.

[4] For more information on U.S. sanctions against Iran, see CRS Report RL32048, *Iran: U.S. Concerns and Policy Responses*, by Kenneth Katzman.

[5] "U.S. Pushes for Tighter United Nations Sanctions Against Iran," *The Oil Daily*, August 7, 2008.

[6]Jonathan S. Landay, "U.S., 5 other nations to seek tougher sanctions against Iran," *The Anniston Star*, August 5, 2008.

[7] NIE, "Iran: Nuclear Intentions and Capabilities," November 2007, accessible at http://www.dni.gov/press_releases / 20071203_release.pdf.

[8] IAEA, "Implementation of the NPT Safeguards Agreement and relevant provisions of Security Council 1737 (2006), 1747 (2007) and 1803 (2008) in the Islamic Republic of Iran," Report by the Director General, May 26, 2008, http://www.iaea.org/Publications/Documents/Board/2008/gov2008-15.pdf.

[9]Iran's fiscal year runs from March 21st to March 20th.

[10]Ibid.

[11]Global Insight, "Iran Country Report," updated December 16, 2008.

[12]IMF, "Regional Economic Outlook: Middle East and Central Asia," *World Economic and Financial Surveys*, October 2008, p. 41.

[13]EIU Country Data.

[14] "Oil falls pile pressure on Tehran," *Middle East Economic Digest*, December 5, 2008.

[15]Anne Penketh, "Iran enters new year in sombre mood as economic crisis bites," *The Independent*, March 24, 2008.

[16]Ibid, p. 21.

[17]CIA, The World Factbook.

[18]EIU, "Iran: Country Profile 2007," p. 33.

[19]About 30% of the population estimated to be under age 15 and less than 5% above age 64 in 2004.

[20]EIU, "Iran: Country Profile 2007," p. 33.

[21]Anne Penketh, "Iran enters new year in sombre mood as economic crisis bites," *The Independent*, March 24, 2008.

[22]Abdelali Jbili, Vitali Kramarenko, and José Bailénm, "Islamic Republic of Iran: Managing the Transition to a Market Economy," IMF, 2007.

[23]EIU, "Iran rank: Macroeconomic risk," January 22, 2008.

[24]Fredrik Dahl, "Iranians worry about high food prices before vote," Reuters, March 6, 2008.

[25] IMF Country Report No. 07/100, "Islamic Republic of Iran: 2006 Article IV Consultation-Staff Report; Staff Statement; Public Information Notice on the Executive Board Discussion; and Statement by the Executive Director for the Islamic Republic of Iran," March 2007, http://www.imf.org/external/ pubs/ft/scr/2007/cr07100.pdf.

[26] Gareth Smyth, "Middle East: Iran cuts farm lending rate in populist 'social justice' move," *Financial Times*, November 15, 2005. Najmeh Bozorgmehr and Roula Khalaf, "World News—Iran: Bank chief takes a realistic tack," *Financial Times*, March 6, 2008.

[27] Najmeh Bozorgmehr and Gareth Smyth, "Coping with the rising cost of marriage, Iranian-style: The new president is to set up a fund to deal with rising expectations of the good life," *Financial Times*, November 8, 2005.

[28]EIU, "Iran: Monetary shrinkage," April 1, 2008.

[29] "Iran: Ahmadi-Nejad populism damages economy," Oxford Analytica, February 19, 2008.

[30]EIU, "Business outlook: Iran," March 1, 2008.

[31]Oxford Analytica, "Prospects 2009: Iran faces economic crisis," November 14, 2008.

[32]Gareth Smyth, "Sanctions fail to fuel dissent on Iran's streets," *Financial Times*, July 24, 2007.

[33]Economic Intelligence Unit (derived from World Bank, World Development Indicators).

[34] "President Makes Damaging Admission on State of Iran's Economy," Global Insight, December 5, 2008.

[35]Oxford Analytica, "Prospects 2009: Iran faces economic crisis," November 18, 2008.

[36] "Iran: Oil Minister Says Iran to Base Budget on Oil Price of $37.50," *Thai News Service*, January 5, 2009.

[37]Nazila Fathi, "Iranian President Proposes Ending Energy Subsidies," *The New York Times*, December 31, 2008.

[38] Global Insight Daily Analysis, "Iran's Bourse Pummelled by Worrying Economic News," December 16, 2008.

[39] "Tehran suffers as finance dries up," *Middle East Economic Digest*, November 1,4 2008.

[40] "Iran: Mostzafan va Janzaban Supports Veterans, Covert Activities," Open Source Center report, May 2, 2006.

[41]EIU, "Iran risk: Legal and regulatory risk," January 22, 2008.

[42] For more information on the IRGC, see Kenneth Katzman, "The Warriors of Islam: Iran's Revolutionary Guard," Westlaw Press, 1993.

[43] Ali Alfoneh, "How Intertwined Are the Revolutionary Guards in Iran's Economy?," American Enterprise Institute, October 22, 2007.

[44]Ibid.
[45]Ibid.
[46]E.O. 13382, "Blocking Property of Weapons of Mass Destruction Proliferators and Their Supporters," June 28, 2005.
[47] Treasury press release, "Statement by Secretary Paulson on Iran Designations," October 25, 2007, http://www.treas.gov/press/releases/hp645.htm. Treasury press release, "Factsheet: Designation of Iranian Entities and Individuals for Proliferation Activities and Support for Terrorism," October 25, 2007, http://www.treas.gov/press/releases/hp644.htm.
[48] "U.S. dilemma: Targeting Iran's oil industry could hurt Iran more," *International Herald Tribune*, November 5, 2007.
[49] Treasury press release, "Factsheet: Designation of Iranian Entities and Individuals for Proliferation Activities and Support for Terrorism," October 25, 2007, http://www.treas.gov/press/releases/hp644.htm.
[50] E.O. 13224, "Blocking Property and Prohibiting Transactions With Persons Who Commit, Threaten to Commit, or Support Terrorism," September 23, 2001.
[51] Treasury press release, "Factsheet: Designation of Iranian Entities and Individuals for Proliferation Activities and Support for Terrorism," October 25, 2007, http://www.treas.gov/press/releases/hp644.htm.
[52]IMF, "Islamic Republic of Iran: 2006 Article IV Consultation ," IMF Country Report No. 07/100, March 2007, p. 12.
[53]Ibid.
[54]Global Insight, "Iran Country Analysis," updated July 10, 2008.
[55]Ibid.
[56] Kenneth Katzman, Specialist in Middle Eastern Affairs, Congressional Research Service, Joint Economic Committee Hearing on Iran, July 25, 2006.
[57]CIA, The World Factbook.
[58]EIU, "Country Profile 2007: Iran," 2007, p. 27.
[59]Ibid, pp. 26-27.
[60]EIA, "Country Analysis Briefs: Iran," October 2007.
[61]Energy Information Administration, "Country Analysis Briefs: Iran," October 2007.
[62]EIU, "Country Profile 2007: Iran," 2007, p. 42.
[63] Samuel Ciszuk, "Tightened Iran Sanctions Introduced by UN Security Council in Anticipation of IAEA Report," February 22, 2008.
[64] "Tehran opens energy sector to overseas investment," *Middle East Economic Digest*, February 8, 2008.
[65]EIU, "Country Profile 2007: Iran," 2007, pp. 35-36.
[66]Javier Blas, "Mideast reels as hunger outgrows oil earnings," *Financial Times*, May 7, 2008.
[67] "Economy & Dutch Disease," *Iran Daily, Islamic Republic News Agency*, April 24, 2007. Lionel Beehner, "What Sanctions Mean for Iran's Economy," Council on Foreign Relations, May 5, 2006.
[68]Global Insights, "Iran Country Analysis," updated July 10, 2008.
[69]EIU, "Country Profile 2007: Iran," 2007, p. 45.
[70]World Steel Association, "World Steel in Figures 2008, 2nd Edition," 2008 http://www.worldsteel.org/.
[71] International Organization of Motor Vehicles (OICA), "World Motor Vehicle Production by Country and Type: 2006-2007," http://oica.net/category/production-statistics/.
[72]Eric Ellis, "Made in Iran," *Fortune Magazine*, September 12, 2006.
[73]Ibid.
[74] EIU, "Country Profile 2007: Iran," 2007, p. 45. Prime Vista Research and Consulting, "Automotive Industry and Marketing of Iran 2007," June 2007.
[75]EIU, "Country Profile 2007: Iran," 2007, p. 46.
[76]Global Insights, "Iran Country Analysis," updated July 10, 2008.
[77]EIU, "Country Profile 2007: Iran," 2007, p. 44.
[78] "Iran Petrochemicals Report Q1 2008," *Business Monitor International*, February 20, 2008.
[79]EIU, "Iran economy: Quick View—Monetary strife—Country Briefing," ViewsWire, April 21, 2008.
[80]Ibid.
[81]Ibid.
[82]EIU, "Iran risk: Financial risk," April 23, 2008.
[83]Global Insight, "Iran Country Analysis," updated July 10, 2008.
[84]Global Insight Daily Analysis, "Iran's Bourse Pummelled by Worrying Economic News," December 16, 2008.
[85] Daniel Glaser, Testimony before the House Committee on Foreign Affairs Subcommittee on the Middle East and South Asia and the Subcommittee on Terrorism, Nonproliferation, and Trade, April 17, 2008, HP-933.
[86] E.O. 13224, "Blocking Property and Prohibiting Transactions With Persons Who Commit, Threaten to Commit, or Support Terrorism," September 23, 2001.
[87]IMF, "Islamic Republic of Iran: 2006 Article IV Consultation ," IMF Country Report No. 07/100, March 2007, p. 17.

[88]E.O. 13382, "Blocking Property of Weapons of Mass Destruction Proliferators and Their Supporters," June 28, 2005.

[89] Treasury press release, "Iran's Bank Sepah Designated By Treasury," January 9, 2007, http://www.treas.gov/ press/ releases/hp219.htm.

[90] Treasury press release, "Statement by Secretary Paulson on Iran Designations," October 25, 2007, http://www.treas.gov/press/releases/hp645.htm. Treasury press release, "Factsheet: Designation of Iranian Entities and Individuals for Proliferation Activities and Support for Terrorism," October 25, 2007, http://www.treas.gov/press/ releases/hp644.htm.

[91] Treasury press release, "Treasury Designates Iran-Controlled Bank for Proliferation: Future Bank Controlled by Iran's Bank Melli," March 12, 2008, http://www.treas.gov/press/releases/hp869.htm.

[92] Treasury press release, "Fact Sheet: Treasury Strengthens Preventative Measures Against Iran," November 6, 2008, http://www.treas.gov/press/releases/hp1258.htm.

[93] "Iran: Iran Dismisses US Allegations against Banks as Ridiculous," *Thai News Service*, August 18, 2008.

[94] "Investment shortfall 'threatens Iran oil output," *AFX Asia*, July 9, 2007.

[95]Steven Lee Myers and Nazila Fathi, "European Leaders Back Bush on Iran," *The New York Times*, June 11, 2008.

[96]Robin Wright, "Iran a Nuclear Threat, Bush Insists; Experts Say President Is Wrong and Is Escalating Tensions," *The Washington Post*, March 22, 2008.

[97] Daniel Dombey and Stephani Kirchgaessner, "Steer clear of Iran central bank, says US," *Financial Times*, March 21, 2008. Jeannine Aversa, "Treasury warns banks that Iran is engaging in deceptive practices to skirt sanctions," Associated Press, March 20, 2008.

[98] Jeannine Aversa, "Treasury warns banks that Iran is engaging in deceptive practices to skirt sanctions," Associated Press, March 20, 2008.

[99]Anna Fifield, "No problem," *Financial Times*, April 14, 2008.

[100] IMF, "Islamic Republic of Iran: 2008 Article IV Consultation," August 2008, IMF Country Report No. 08/284.

[101] Ibid.

[102] Ibid.

[102] Energy Information Administration (EIA), "World Transit Oil Chokepoints," January 2008.

[103] IMF, "Islamic Republic of Iran: 2006 Article IV Consultation ," IMF Country Report No. 07/100, March 2007, p. 29.

[104] Energy Information Administration (EIA), "Country Analysis Briefs: Iran," October 2007.

[105] Telephone conversation with EIA official, April 29, 2008.

[106] "Reports of Indian Company's Decision to Halt Gas Shipments to Iran Welcomed by Sherman," International Trade Daily, January 8, 2009. "Rep. Sherman Issues Statement on Reported RIL Decision to Stop Selling Refined Petroleum to Iran," US Fed News Service, January 7, 2009.

[107] Export-Import Bank of the United States, *Annual Report 2008* and *Annual Report 2007*, Washington, DC.

[108] Power and Interest News Report (PINR), "Iran Looks for Allies through Asian and Latin American Partnerships," May 27, 2007.

[109] International Monetary Fund (IMF), Direction of Trade Statistics.

[110] Global Insight, "Iran Country Report," updated July 10, 2008.

[111] "Tajik speaker says Tajikistan keen on active cooperation with Iran," ASIA-Plus Information Agency, March 27, 2008.

[112] "Iran, Azerbaijan, Turkey to build joint railway," BBC Monitoring Caucasus, March 27, 2008.

[113] Ibid.

[114] Barbara Surk, "Dubai at center of US efforts to pressure Iran, "Associated Press Newswires, October 26, 2007.

[115] "UAE/Iran: Trade squeeze, or business as usual?", EIU—Business Middle East, January 16, 2008.

[116] Global Insight, Country Intelligence—Analysis, "U.A.E. Banks Latest to Restrict Iran's Access to Credit," October 28, 2008.

[117] Ibid.

[118] Ibid.

[119] GAO-08-58, "Iran Sanctions: Impact in Furthering U.S. Objectives Is Unclear and Should Be Reviewed," p. 18, December 2007.

[120] Jeffrey J. Schott, "The Iran and Libya Sanctions Act of 1996: Results to Date," Peterson Institute for International Economics, Testimony before the Committee on International Relations, U.S. House of Representatives, July 23, 1997.

[121] Glenn R. Simpson, "Democrats Urge Sanctions on Iran's Central Bank," *The Wall Street Journal*, March 5, 2008.

[122] Danielle Pletka, "Congress's Ill-Timed Iran Bills," *Washington Post*, August 28, 2007.

[123] Ibid.

[124] Bertrand Benoit, "Berlin hardens trade stance with Iran," *Financial Times*, February 11, 2008.

[125] "German imports from Iran up despite nuclear row-paper," *Reuters News*, January 8, 2008.

[126] Samuel Ciszuk, "UN Security Council Tightens Iran Sanctions, Complicating Oil and Gas Developments and Trade," Global Insight Daily Analysis, March 4, 2008.

[127] Anna Fifield, "No problem," *Financial Times*, April 14, 2008.
[128] Global Insight, Country Intelligence—Analysis, "Iran: U.S. and UN Sanctions Hit Iran's Shipping Industry," April 29, 2008.
[129] IMF, "Islamic Republic of Iran: 2006 Article IV Consultation ," IMF Country Report No. 07/100, March 2007, p. 18.
[130] Fiona Fleck, "Iraq Is Granted Observer Status at the WTO," *The New York Times*, February 12, 2004.
[131] IMF, "Islamic Republic of Iran: 2006 Article IV Consultation ," IMF Country Report No. 07/100, March 2007, p. 18.
[132] IMF, "Islamic Republic of Iran: 2006 Article IV Consultation ," IMF Country Report No. 08/284, August 2008.
[133] Global Insights, "Iran Country Analysis," updated March 30, 2009.
[134] "Iran stops accepting U.S. dollars for oil," RIA Novosti, December 8, 2007.
[135] Najmeh Bozorgmehr and Roula Khalaf, "World News—Iran: Bank chief takes a realistic tack," *Financial Times*, March 6, 2008.
[136] "Iranian oil no longer available for U.S. dollars," RIA Novosti, December 11, 2007.
[137] United Nations Conference on Trade and Development (UNCTAD), Foreign Direct Investment online database.
[138] IMF, "Islamic Republic of Iran: 2006 Article IV Consultation ," IMF Country Report No. 07/100, March 2007, p. 29.
[139] See CRS Report RS20871, *The Iran Sanctions Act (ISA)*, by Kenneth Katzman.
[140] "Iran economy: Oil breakthrough?," Economist Intelligence Unit, December 10, 2007.
[141] Global Insight, "Iran Country Report," updated July 10, 2008.
[142] BMI Industry Insights, "StatoilHydro Pulling Out of Iran, But Pulling In The Profits," August 1, 2008.
[143] Eli Lake, "State Department Looks to Sanction Law," *The New York Sun*, March 21, 2008.
[144] "Iran: Sanctions and threats damage economy," Oxford Analytica, June 15, 2007.
[145] GAO-08-58, "Iran Sanctions: Impact in Furthering U.S. Objectives Is Unclear and Should Be Reviewed," p. 18, December 2007.
[146] EIU, "Iran economy: Au revoir LNG?—Country Briefing," ViewsWire, May 12, 2008.
[147] "Tehran opens energy sector to overseas investment," *Middle East Economic Digest*, February 8, 2008.
[148] "Divesting from Iran: State-by-State Update," Israel Project news release, accessed via US Fed News, February 21, 2008.
[149] Benjamin Weinthal, "Switzerland to sign second-largest Iran gas deal today," *The Jerusalem Post*, March 17, 2008.
[150] "Gazprom announces gas deals with Iran and Nigeria," *Energy Economist*, March 1, 2008.
[151] "Chinese delegation to sign major gas deal soon: source," Platts Commodity News, March 5, 2008.
[152] David Winning and Renya Peng, "Update: CNOOC, Iran to Ink Deal for 10 Mln Tons LNG-Sources," Dow Jones Newswires, January 21, 2008.
[153] "Iran, Turkey to Discuss Gas Projects," *Turkish Daily News*, May 5, 2008. "Iran-Turkey Deals Could Prove Key to Europe's Gas Import Plans," *International Oil Daily*, December 1, 2008.
[154] Rikard Jozwiak, "EU exports to Iran rise," *European Voice*, July 8, 2008.
[155] "EU exports to Iran rising despite sanctions," Reuters EU Highlights, August 6, 2008.
[156] "Iran/Turkey: Iran, Turkey Plan to Set up Gas Company," *Thai News Service*, December 29, 2008.
[157] Oxford Analytica, "Gas deal points to Bahraini confidence," Global Strategic Analysis, Monday 8, 2008.
[158] Oxford Analytica, "Russia/Iran: Swap agreement bolsters energy ties," April 8, 2009.
[159] World Bank data accessed August 20, 2008.
[160] For more information on World Bank lending to Iran, see CRS Report RS22704, *The World Bank and Iran*, by Martin A. Weiss and Jonathan E. Sanford.
[161] Jay Solomon, "U.S. Sanctions Bahrain Bank Over Iran," *The Wall Street Journal*, March 13, 2008.
[162] Jeffrey J. Schott, "The Iran and Libya Sanctions Act of 1996: Results to Date," Peterson Institute of International Economics, Testimony before the Committee on International Relations, U.S. House of Representatives, July 23, 1997.
[163] William A. Reinsch, President of National Foreign Trade Council and Co-Chairman of USA*Engage, Testimony before the U.S. Senate Committee on Finance, Regarding S. 970, The Iran Counter-Proliferation Act of 2007," April 8, 2008.
[164] "Iran: Sanctions and threats damage economy," Oxford Analytica, June 15, 2007.
[165] GAO-08-58, "Iran Sanctions: Impact in Furthering U.S. Objectives Is Unclear and Should Be Reviewed," p. 18, December 2007.
[166] Adam Graham-Silverman, "Despite Flurry of Action in House, Congress Unlikely to Act Against Iran," *Congressional Quarterly Today*, September 12, 2007.
[167] Jad Mouawad, "Iran feels West's grip; Energy troubles carry risks of protests," *International Herald Tribune*, February 13, 2007.

In: Iran Social, Economic and Political Developments
Editor: Joshua A. Owenstein
ISBN: 978-1-60741-552-7

Chapter 6

IRAN'S NUCLEAR PROGRAM: TEHRAN'S COMPLIANCE WITH INTERNATIONAL OBLIGATIONS*

Paul K. Kerr

SUMMARY

In 2002, the International Atomic Energy Agency (IAEA) began investigating allegations that Iran had conducted clandestine nuclear activities; the agency ultimately reported that some of these activities had violated Tehran's IAEA safeguards agreement. The agency has not stated definitively that Iran has pursued nuclear weapons, but has also not yet been able to conclude that the country's nuclear program is exclusively for peaceful purposes. The IAEA Board of Governors referred the matter to the U.N. Security Council in February 2006. Since then, the council has adopted five resolutions, the most recent of which (Resolution 1835) was adopted in September.

The Security Council has required Iran to cooperate fully with the IAEA's investigation of its nuclear activities, suspend its uranium enrichment program, suspend its construction of a heavy- water reactor and related projects, and ratify the Additional Protocol to its IAEA safeguards agreement. However, a February 2009 report from IAEA Director-General Mohamed ElBaradei to the agency's Board of Governors indicated that Tehran has continued to defy the council's demands by continuing work on its uranium enrichment program and heavy-water reactor program. Iran has signed, but not ratified, its Additional Protocol.

Iran and the IAEA agreed in August 2007 on a work plan to clarify the outstanding questions regarding Tehran's nuclear program. Most of these questions have essentially been resolved, but ElBaradei told the agency's board in June 2008 that the agency still has questions regarding "possible military dimensions to Iran's nuclear programme." According to ElBaradei's report, the IAEA has not been able to make progress on these matters.

---

* This is an edited, reformatted and augmented version of a CRS Report for Congress publication dated March 2009.

This chapter provides a brief overview of Iran's nuclear program and describes the legal basis for the actions taken by the IAEA board and the Security Council.

INTRODUCTION

Iran ratified the nuclear Nonproliferation Treaty (NPT) in 1970. Article III of the treaty requires non-nuclear-weapon states-parties[1] to accept comprehensive International Atomic Energy Agency (IAEA) safeguards; Tehran concluded a comprehensive safeguards agreement with the IAEA in 1974.[2] In 2002, the agency began investigating allegations that Iran had conducted clandestine nuclear activities; the IAEA ultimately reported that some of these activities had violated Tehran's safeguards agreement. The agency has not stated definitively that Iran has pursued nuclear weapons, but has also not yet been able to conclude that the country's nuclear program is exclusively for peaceful purposes. The IAEA continues to investigate the program.

Following more than three years of investigation, the IAEA Board of Governors referred the matter to the U.N. Security Council in February 2006. Since then, the council has adopted five resolutions requiring Iran to take steps to alleviate international concerns about its nuclear program. This chapter provides a brief overview of Iran's nuclear program and describes the legal basis for the actions taken by the IAEA board and the Security Council.

For more detailed information about Iran's nuclear program, see CRS Report RL34544, *Iran's Nuclear Program: Status*, by Paul K. Kerr. For more information about the state of international diplomacy with Iran, see CRS Report RL32048, *Iran: U.S. Concerns and Policy Responses*, by Kenneth Katzman.

Background

Iran's construction of a gas centrifuge-based uranium enrichment facility is currently the main source of proliferation concern. Gas centrifuges enrich uranium by spinning uranium hexafluoride gas at high speeds to increase the concentration of the uranium-235 isotope. Such centrifuges can produce both low-enriched uranium (LEU), which can be used in nuclear power reactors, and highly enriched uranium (HEU), which is one of the two types of fissile material used in nuclear weapons. HEU can also be used as fuel in certain types of nuclear reactors. Iran also has a uranium-conversion facility, which converts uranium oxide into several compounds, including uranium hexafluoride. Tehran claims that it wants to produce LEU for its current and future power reactors.

Iran's construction of a reactor moderated by heavy water has also been a source of concern. Although Tehran says that the reactor is intended for the production of medical isotopes, it is a proliferation concern because the reactor's spent fuel will contain plutonium well-suited for use in nuclear weapons. In order to be used in nuclear weapons, however, plutonium must be separated from the spent fuel—a procedure called "reprocessing." Iran has said that it will not engage in reprocessing.

Iran and the IAEA agreed in August 2007 on a work plan to clarify the outstanding questions regarding Tehran's nuclear program.[3] Most of these questions, which had created suspicions that Iran had been pursuing a nuclear weapons program, have essentially been resolved. IAEA Director-General Mohamed ElBaradei, however, told the IAEA board June 2, 2008, that there is "one remaining major [unresolved] issue," which concerns questions regarding "possible military dimensions to Iran's nuclear programme." A February 19, 2009, report[4] from ElBaradei to the Security Council and the IAEA board states that the agency "has not made any substantive progress" on these matters. Tehran has questioned the authenticity of some of the evidence underlying the agency's concerns and maintains that it has not done any work on nuclear weapons.

The government has also expressed concern to the IAEA that resolving some of these issues would require agency inspectors to have "access to sensitive information related to its conventional military and missile related activities." The IAEA, according to a September 2008 report from ElBaradei, has stated its willingness to discuss with Iran

> modalities that could enable Iran to demonstrate credibly that the activities referred to in the documentation are not nuclear related, as Iran asserts, while protecting sensitive information related to its conventional military activities.[5]

Indeed, the agency says that it has made several specific proposals, but Tehran has not yet provided the requested information.

The most recent Security Council resolution (1835), adopted September 27, 2008, requires Iran to cooperate fully with the IAEA's investigation of its nuclear activities, suspend its uranium enrichment program, suspend its construction of a heavy-water reactor and related projects, and ratify the Additional Protocol to its IAEA safeguards agreement.[6] However, ElBaradei's September report indicated that Tehran has continued to defy the council's demands by continuing work on its uranium enrichment program and heavy-water reactor program. Iranian officials have repeatedly stated that Iran will not suspend its enrichment program. Tehran has signed, but not ratified, its Additional Protocol.

Iran and the IAEA

As noted above, Iran is a party to the NPT and has concluded a comprehensive safeguards agreement. According to the IAEA, safeguards pursuant to such agreements

> are applied to verify a State's compliance with its undertaking to accept safeguards on all nuclear material in all its peaceful nuclear activities and to verify that such material is not diverted to nuclear weapons or other nuclear explosive devices.[7]

Comprehensive safeguards are designed to enable the IAEA to detect the diversion of nuclear material from peaceful purposes to nuclear weapons uses, as well as to detect undeclared nuclear activities and material.[8] Safeguards include agency inspections and monitoring of declared nuclear facilities.

The agency's inspections and monitoring authority in a particular country are limited to facilities that have been declared by the government. Additional Protocols to IAEA comprehensive safeguards agreements increase the agency's ability to investigate clandestine nuclear facilities and activities by increasing the IAEA's authority to inspect certain nuclear-related facilities and demand information from member-states.[9] Iran signed such a protocol in December 2003 and agreed to implement the agreement pending ratification. Tehran stopped adhering to its Additional Protocol in 2006.

The IAEA's authority to investigate nuclear-weapons-related activity is limited. ElBaradei explained in a 2005 interview that "... we don't have an all-encompassing mandate to look for every computer study on weaponization. Our mandate is to make sure that all nuclear materials in a country are declared to us."[10] Similarly, a February 2006 report from ElBaradei to the IAEA board stated that "absent some nexus to nuclear material the Agency's legal authority to pursue the verification of possible nuclear weapons related activity is limited."[11]

The current public controversy over Iran's nuclear program began in August 2002, when the National Council of Resistance on Iran (NCRI), an Iranian exile group, revealed information during a press conference (some of which later proved to be accurate) that Tehran had built nuclear-related facilities that it had not revealed to the IAEA. The United States had been aware of at least some of these activities, according to knowledgeable former officials.[12] Prior to the NCRI's revelations, the IAEA had expressed concerns that Iran had not been providing the agency with all relevant information about its nuclear programs, but had never found the country in violation of its safeguards agreement.

In fall 2002, the IAEA began to investigate Iran's nuclear activities at the sites named by the NCRI; inspectors visited the sites the following February. Adopting its first resolution (GOV/2003/69)[13] on the matter in September 2003, the IAEA board called on Tehran to increase its cooperation with the agency's investigation, suspend its uranium enrichment activities, and "unconditionally sign, ratify and fully implement" an Additional Protocol.

In October 2003, Iran concluded a voluntary agreement with France, Germany, and the United Kingdom, collectively known as the "E3," to suspend its enrichment activities, sign and implement an Additional Protocol to its IAEA safeguards agreement, and comply fully with the IAEA's investigation.[14] As a result, the agency's board decided to refrain from referring the matter to the U.N. Security Council. As noted above, Tehran signed this Additional Protocol in December 2003, but has never ratified it.

Ultimately, the IAEA's investigation, as well as information Iran provided after the October 2003 agreement, revealed that Iran had engaged in a variety of clandestine nuclear-related activities, some of which violated the country's safeguards agreement (see Appendix A). After October 2003, Iran continued some of its enrichment-related activities, but Tehran and the E3 agreed in November 2004 to a more detailed suspension agreement.[15] However, Iran resumed uranium conversion in August 2005 under the leadership of President Mahmoud Ahmadinejad, who had been elected two months earlier.

On September 24, 2005, the IAEA Board of Governors adopted a resolution (GOV/2005/77)[16] that, for the first time, found Iran to be in noncompliance with its IAEA safeguards agreement. The board, however, did not refer Iran to the Security Council, choosing instead to give Tehran additional time to comply with the board's demands. The resolution urged Iran:

- To implement transparency measures... [including] access to individuals, documentation relating to procurement, dual use equipment, certain military owned workshops and research and development locations;
- To re-establish full and sustained suspension of all enrichment-related activity;

- To reconsider the construction of [the] research reactor moderated by heavy water;
- To ratify promptly and implement in full the Additional Protocol;
- [T]o continue to act in accordance with the provisions of the Additional Protocol.

No international legal obligations required Tehran to take these steps. But ElBaradei's September 2008 report asserted that, without Iranian implementation of such "transparency measures," the IAEA "will not be in a position to progress in its verification of the absence of undeclared nuclear material and activities in Iran."

Iran announced in January 2006 that it would resume research and development on its centrifuges at Natanz. In response, the IAEA board adopted a resolution (GOV/2006/14)[17] February 4, 2006, referring the matter to the Security Council and reiterating its call for Iran to take the measures specified in the September resolution. Two days later, Tehran announced that it would stop implementing its Additional Protocol.

Iran further scaled back its cooperation with the IAEA in March 2007, when the government decided to stop complying with a portion of the subsidiary arrangements for its IAEA safeguards agreement.[18] That provision, to which Iran agreed in 2003, requires Tehran to provide design information for new nuclear facilities "as soon as the decision to construct, or to authorize construction, of such a facility has been taken, whichever is earlier." Previously, Iran was required to provide design information for a new facility only 180 days before introducing nuclear material into it.

The IAEA has asked Iran to "reconsider" this decision, which appears to be inconsistent with Tehran's obligations under its subsidiary arrangements. Article 39 of Iran's safeguards agreement states that those arrangements "may be extended or changed by agreement between" Iran and the IAEA, but does not provide for a unilateral modification or suspension of any portion of those arrangements.[19] Moreover, a March 2009 statement[20] by the IAEA legal adviser states that Tehran's actions subsequent to its March 2007 decision are "inconsistent with" Iran's obligations under its subsidiary arrangements. Specifically, Tehran has failed to provide the IAEA with design information for a reactors that, according to Iranian officials, Iran intends to construct at Darkhovin. Similarly, Tehran has not provided updated design information for the heavy water reactor that is under construction.

The IAEA legal adviser's statement says that "it is difficult to conclude that" Iran's decision to stop complying with the updated provision of its subsidiary arrangements "in itself constitutes non-compliance with, or a breach of" Tehran's safeguards agreement. However, the statement does characterize Iran's ongoing refusal to allow IAEA inspectors to carry out Design Information Verification of the heavy water reactor as "inconsistent with" Tehran's obligations under its safeguards agreement.[21]

Iran and the U.N. Security Council

On March 29, 2006, the U.N. Security Council President issued a statement that called on Iran to "take the steps required" by the February IAEA board resolution. Such statements are, however, not legally binding. The council subsequently adopted five resolutions concerning Iran's nuclear program: 1696 (July 2006), 1737 (December 2006), 1747 (March 2007), 1803 (March 2008), and 1835 (September 2008). The second, third, and fourth resolutions imposed a variety of restrictions on Iran.

Resolution 1696 was the first to place legally binding Security Council requirements on Iran with respect to its nuclear program. That resolution made mandatory the IAEA—demanded suspension and called on Tehran to implement the transparency measures called for by the IAEA board's February 2006 resolution. Resolution 1737 reiterated these requirements but expanded the suspension's scope to include "work on all heavy water-related projects."

It is worth noting that the Security Council has acknowledged (in Resolution 1803, for example) Iran's rights under Article IV of the NPT, which states that parties to the treaty have "the inalienable right ... to develop research, production and use of nuclear energy for peaceful Purposes."[22]

AUTHORITY FOR IAEA AND U.N. SECURITYCOUNCIL ACTIONS

The legal authority for the actions taken by the IAEA Board of Governors and the U.N. Security Council is found in both the IAEA Statute and the U.N. Charter. The following sections discuss the relevant portions of those documents.

IAEA Statute

Two sections of the IAEA Statute explain what the agency should do if an IAEA member-state is found to be in noncompliance with its safeguards agreement.[23] Article III B. 4. of the statute states that the IAEA is to submit annual reports to the U.N. General Assembly and, "when appropriate," to the U.N. Security Council. If "there should arise questions that are within the competence of the Security Council," the article adds, the IAEA "shall notify the Security Council, as the organ bearing the main responsibility for the maintenance of international peace and security."

Additionally, Article XII C. states that IAEA inspectors are to report non-compliance issues to the agency's Director-General, who is to report the matter to the IAEA Board of Governors. The board is then to "call upon the recipient State or States to remedy forthwith any non-compliance which it finds to have occurred," as well as "report the non-compliance to all members and to the Security Council and General Assembly of the United Nations."

In the case of Iran, the September 24, 2005 IAEA board resolution (GOV/2005/77) stated that the board

> found that Iran's many failures and breaches of its obligations to comply with its NPT Safeguards Agreement, as detailed in GOV/2003/75 [a November 2003 report from ElBaradei], constitute non compliance in the context of Article XII.C of the Agency's Statute;

According to the resolution, the board also found

> that the history of concealment of Iran's nuclear activities referred to in the Director General's report [GOV/2003/75], the nature of these activities, issues brought to light in the course of the Agency's verification of declarations made by Iran since September 2002 and the resulting absence of confidence that Iran's nuclear programme is exclusively for peaceful

> purposes have given rise to questions that are within the competence of the Security Council, as the organ bearing the main responsibility for the maintenance of international peace and security.

ElBaradei issued the report cited by the resolution, GOV/2003/75, in November 2003.[24] It described a variety of Iranian nuclear activities, which are detailed in Appendix A, that violated Tehran's safeguards agreement. ElBaradei has since reported that Iran has taken corrective measures to address these safeguards breaches. As noted above, the 2005 resolution called on Iran to take a variety of actions that Tehran was not legally required to implement.

U.N. Charter and the Security Council

Several articles of the U.N. Charter, which is a treaty, describe the Security Council's authority to impose requirements and sanctions on Iran.[25] Article 24 confers on the council "primary responsibility for the maintenance of international peace and security." The article also states that the "specific powers granted to the Security Council for the discharge of these duties are laid down" in several chapters of the charter, including Chapter VII, which describes the actions that the council may take in response to "threats to the peace, breaches of the peace, and acts of aggression."

Chapter VII of the charter contains three articles relevant to the Iran case. Security Council resolutions that made mandatory the IAEA's demands concerning Iran's nuclear program invoked Chapter VII. Article 39 of that chapter states that the council

> shall determine the existence of any threat to the peace, breach of the peace, or act of aggression and shall make recommendations, or decide what measures shall be taken in accordance with Articles 41 and 42, to maintain or restore international peace and security.

Resolution 1696 invoked Article 40 of Chapter VII "in order to make mandatory the suspension required by the IAEA." As noted earlier, that resolution did not impose any sanctions on Iran. Article 40 states that

> the Security Council may, before making the recommendations or deciding upon the measures provided for in Article 39 [of Chapter VII], call upon the parties concerned to comply with such provisional measures as it deems necessary or desirable.

Resolutions 1737, 1747, and 1803, which did impose sanctions, invoked Article 41 of Chapter VII. According to Article 41, the Security Council

> may decide what measures not involving the use of armed force are to be employed to give effect to its decisions, and it may call upon the Members of the United Nations to apply such measures. These may include complete or partial interruption of economic relations and of rail, sea, air, postal, telegraphic, radio, and other means of communication, and the severance of diplomatic relations.

As noted above, Security Council resolution 1835 did not impose new sanctions, but reaffirmed the previous resolutions and called on Iran to comply with them. The sanctions imposed by those resolutions remain in place. The five permanent members of the council, along with Germany, are continuing to discuss the matter.

It is worth noting that Article 25 of the U.N. Charter obligates U.N. members "to accept and carry out the decisions of the Security Council."

HAS IRAN VIOLATED THE NPT?[26]

Whether Iran has violated the NPT is unclear. The treaty does not contain a mechanism for determining that a state-party has violated its obligations. Moreover, there does not appear to be a formal procedure for determining such violations. An NPT Review Conference would, however, be one venue for NPT states-parties to make such a determination.

The U.N. Security Council has never declared Iran to be in violation of the NPT; neither the council nor the U.N. General Assembly has a responsibility to adjudicate treaty violations. However, the lack of a ruling by the council on Iran's compliance with the NPT has apparently had little practical effect because, as noted above, the council has taken action in response to the IAEA Board of Governors' determination that Iran has violated its safeguards agreement.

Iran's violations of its safeguards agreement appear to constitute violations of Article III, which requires NPT non-nuclear-weapon states-parties to adhere to their safeguards agreements. Tehran may also have violated provisions of Article II which state that non-nuclear-weapon states-parties shall not "manufacture or otherwise acquire nuclear weapons or other nuclear explosive devices" or "seek or receive any assistance in the manufacture of nuclear weapons or other nuclear explosive devices."

As previously noted, the IAEA is continuing to investigate evidence of what ElBaradei described in June 2008 as "possible military dimensions to Iran's nuclear programme." Such activities may indicate that Tehran has violated both Article II provisions described above, but the IAEA has never reported that Iran has attempted to develop nuclear weapons.

Despite the lack of such an IAEA conclusion, a 2005 State Department report argued that the country had violated Article II of the NPT:

> The breadth of Iran's nuclear development efforts, the secrecy and deceptions with which they have been conducted for nearly 20 years, its redundant and surreptitious procurement channels, Iran's persistent failure to comply with its obligations to report to the IAEA and to apply safeguards to such activities, and the lack of a reasonable economic justification for this program leads us to conclude that Iran is pursuing an effort to manufacture nuclear weapons, and has sought and received assistance in this effort in violation of Article II of the NPT.[27]

The report also stated that Iran's "weapons program combines elements" of Tehran's declared nuclear activities, as well as suspected "undeclared fuel cycle and other activities that may exist, including those that may be run solely be the military."

The State Department's reasoning appears to be based on an interpretation of the NPT which holds that a wide scope of nuclear activities could constitute violations of Article II. The 2005 report states that assessments regarding Article II compliance "must look at the totality of the facts, including judgments as to" a state-party's "purpose in undertaking the nuclear activities in question." The report also includes a list of activities which could constitute such noncompliance.[28]

The 2005 State Department report cites testimony from then-Arms Control and Disarmament Agency Director William Foster during a 1968 Senate Foreign Relations Committee hearing.[29] Foster stated that "facts indicating that the purpose of a particular activity was the acquisition of a nuclear explosive device would tend to show non-compliance" with Article II. He gave two examples: "the construction of an experimental or prototype nuclear explosive device" and "the production of components which could only have relevance" to such a device. However, Foster also noted that a variety of other activities could also violate Article II, adding that the United States believed it impossible "to formulate a comprehensive definition or interpretation."

A November 2007 National Intelligence Estimate stated that "until fall 2003, Iranian military entities were working under government direction to develop nuclear weapons."[30] This past program could be a violation of Article II, although the estimate does not provide any detail about the program.

It is worth noting that the State Department's arguments appear to rely heavily on the notion that a state's apparent intentions underlying certain nuclear-related activities can be used to determine violations to Article II. This interpretation is not shared by all experts.[31]

Appendix A. Iranian Noncompliance with Its IAEA Safeguards Agreement

The November 2003 report (GOV/2003/75) from IAEA Director-General ElBaradei to the agency's Board of Governors details what the September 2005 board resolution described as "Iran's many failures and breaches of its obligations to comply with its safeguards agreement."

The report stated that

> Iran has failed in a number of instances over an extended period of time to meet its obligations under its Safeguards Agreement with respect to the reporting of nuclear material and its processing and use, as well as the declaration of facilities where such material has been processed and stored.

The report detailed some of these failures and referenced other failures described in two earlier reports (GOV/2003/40 and GOV/2003/63) from ElBaradei to the IAEA board.[32]

According to GOV/2003/40, Iran failed to declare the following activities to the agency:

- The importation of natural uranium, and its subsequent transfer for further processing.
- The processing and use of the imported natural uranium, including the production and loss of nuclear material, and the production and transfer of resulting waste.

Additionally, Iran failed to

- declare the facilities where nuclear material (including the waste) was received, stored and processed,
- provide in a timely manner updated design information for a research reactor located in Tehran, as well as
- provide in a timely manner information on two waste storage sites.

GOV/2003/63 stated that Iran failed to report uranium conversion experiments to the IAEA. According to GOV/2003/75, Iran failed to report the following activities to the IAEA:

- The use of imported natural uranium hexafluoride for the testing of centrifuges, as well as the subsequent production of enriched and depleted uranium.
- The importation of natural uranium metal and its subsequent transfer for use in laser enrichment experiments, including the production of enriched uranium, the loss of nuclear material during these operations, and the production and transfer of resulting waste.
- The production of a variety of nuclear compounds from several different imported nuclear materials, and the production and transfer of resulting wastes.
- The production of uranium targets and their irradiation in the Tehran Research Reactor, the subsequent processing of those targets (including the separation of plutonium), the production and transfer of resulting waste, and the storage of unprocessed irradiated targets.

Iran also failed to provide the agency with design information for a variety of nuclear-related facilities, according to the report. These included the following:

- A centrifuge testing facility.
- Two laser laboratories and locations where resulting wastes were processed.
- Facilities involved in the production of a variety of nuclear compounds.
- The Tehran Research Reactor (with respect to the irradiation of uranium targets), the hot cell facility where the plutonium separation took place, as well as the relevant waste handling facility.

Additionally, the report cited Iran's "failure on many occasions to co-operate to facilitate the implementation of safeguards, through concealment" of its nuclear activities.

Appendix B. Extended Remarks by William FosterRegarding Possible NPT Article II Violations

On July 10, 1968, then-Arms Control and Disarmament Agency Director William Foster testified before the Senate Foreign Relations Committee about the NPT. In response to a question regarding the type of nuclear activities prohibited by Article II of the treaty, Foster supplied the following statement:

Extension of Remarks by Mr. Foster in Response to Question Regarding Nuclear Explosive Devices

The treaty articles in question are Article II, in which non-nuclear-weapon parties undertake "not to manufacture or otherwise acquire nuclear weapons or other nuclear explosive devices," and Article IV, which provides that nothing in the Treaty is to be interpreted as affecting the right of all Parties to the Treaty "to develop research, production and use of nuclear energy for peaceful purposes...in conformity with Articles I and II of this Treaty." In the course of the negotiation of the Treaty, United States representatives were asked their views on what would constitute the "manufacture" of a nuclear weapon or other nuclear explosive device under Article II of the draft treaty. Our reply was as follows:

"While the general intent of this provision seems clear, and its application to cases such as those discussed below should present little difficulty, the United States believe [sic] it is not possible at this time to formulate a comprehensive definition or interpretation. There are many hypothetical situations which might be imagined and it is doubtful that any general definition or interpretation, unrelated to specific fact situations could satisfactorily deal with all such situations."

"Some general observations can be made with respect to the question of whether or not a specific activity constitutes prohibited manufacture under the proposed treaty. For example, facts indicating that the purpose of a particular activity was the acquisition of a nuclear explosive device would tend to show non-compliance. (Thus, the construction of an experimental or prototype nuclear explosive device would be covered by the term 'manufacture' as would be the production of components which could only have relevance to a nuclear explosive device.) Again, while the placing of a particular activity under safeguards would not, in and of itself, settle the question of whether that activity was in compliance with the treaty, it would of course be helpful in allaying any suspicion of non-compliance."

"It may be useful to point out, for illustrative purposes, several activities which the United States would not consider per se to be violations of the prohibitions in Article II. Neither uranium enrichment nor the stockpiling of fissionable material in connection with a peaceful program would violate Article II so long as these activities were safeguarded under Article III. Also clearly permitted would be the development, under safeguards, of plutonium fueled power reactors, including research on the properties of metallic plutonium, nor would Article II interfere with the development or use of fast breeder reactors under safeguards."

End Notes

1 Under the NPT, the five nuclear-weapon states are China, France, Russia, the United Kingdom, and the United States. All other parties are non-nuclear-weapon states.

2 INFCIRC/214, available at http://www.iaea.org/Publications/Documents/Infcircs/Others/infcirc214.pdf. Non-nuclear-weapons states-parties to the nuclear Nonproliferation Treaty are obligated to conclude comprehensive safeguards agreements with the International Atomic Energy Agency.

3 The text is available at http://www.iaea.org/Publications/Documents/Infcircs/2007/infcirc711.pdf.

4 GOV/2009/8, available at http://www.iaea.org/Publications/Documents/Board/2009/gov2009-8.pdf.

5 GOV/2008/38, available at http://www.iaea.org/Publications/Documents/Board/2008/gov2008-38.pdf.

6 Iran is also constructing a plant for the production of heavy water.

7 IAEA Safeguards Glossary, op cit. Comprehensive safeguards agreements are based on a model described in INFCIRC 153, available at http://www.iaea.org/Publications/Documents/Infcircs/Others/infcirc153.pdf.

8 Ibid.

[9] Additional Protocols for an individual IAEA member-state are based on the agency's Model Additional Protocol (INFCIRC/540), available at http://www.iaea.org/Publications/Documents/Infcircs/1997/infcirc540c.pdf.

[10] "Tackling the Nuclear Dilemma: An Interview with IAEA Director-General Mohamed ElBaradei," February 4, 2005, available at http://www.armscontrol.org/act/2005_03/ElBaradei.

[11] GOV/2006/15, available at http://www.iaea.org/Publications/Documents/Board/2006/gov2006-15.pdf.

[12] Gary Samore, Former Senior Director for Nonproliferation and Export Controls on the National Security Council, personal communication June 5, 2008; Director of Central Intelligence George J. Tenet, "DCI Remarks on Iraq's WMD Programs," February 5, 2004, available at https://www.cia.gov/news-information/speeches-testimony/2004/ tenet_georgetownspeech_02052004.html.

[13] Available at http://www.iaea.org/Publications/Documents/Board/2003/gov2003-69.pdf.

[14] The text of the agreement is available at http://www.iaea.org/NewsCenter/Focus/IaeaIran/ statement_iran2110 2003.shtml.

[15] The text of the agreement is available at http://www.iaea.org/Publications/ Documents/ Infcircs/ 2004/infcirc637.pdf.

[16] Available at http://www.iaea.org/Publications/Documents/Board/2005/gov2005-77.pdf.

[17] Available at http://www.iaea.org/Publications/Documents/Board/2006/gov2006-14.pdf.

[18] According to the 2001 IAEA Safeguards Glossary, subsidiary arrangements describe the "technical and administrative procedures for specifying how the provisions laid down in a safeguards agreement are to be applied."

[19] See also GOV/2007/22, available at http://www.iaea.org/Publications/Documents/Board/2007/gov2007-22.pdf.

[20] Available at http://www.armscontrolwonk.com/file_download/162/Legal_Adviser_Iran.pdf.

[21] Iran stated in an April 2007 letter to the IAEA that, given Tehran's March 2007 decision regarding the subsidiary arrangements to its safeguards agreement, such visits were unjustified.

[22] The treaty text is available at http://www.iaea.org/Publications/Documents/Infcircs/Others/infcirc140.pdf.

[23] The text of the IAEA Statute is available at http://www.iaea.org/About/statute_text.html.

[24] Available at http://www.iaea.org/Publications/Documents/Board/2003/gov2003-75.pdf.

[25] The text of the charter is available at http://www.un.org/aboutun/charter/.

[26] Portions of this section are based on interviews with U.N. and State Department officials.

[27] *Adherence to and Compliance with Arms Control, Nonproliferation and Disarmament Agreements and Commitments*, Department of State, August 2005, available at http://www.state.gov/documents/ organization/52113.pdf.

[28] According to the report, such activities can include (1) the presence of undeclared nuclear facilities; (2) procurement patterns inconsistent with a civil nuclear program (e.g., clandestine procurement networks, possibly including the use of front companies, false end-use information, and fraudulent documentation); (3) security measures beyond what would be appropriate for peaceful, civil nuclear installations; (4) a pattern of Article III safeguards violations suggestive not of mere mistake or incompetence, but of willful violation and/or systematic deception and denial efforts aimed at concealing nuclear activities from the IAEA; and (5) a nuclear program with little (or no) coherence for peaceful purposes, but great coherence for weapons purposes.

[29] *Nonproliferation Treaty*, Senate Committee on Foreign Relations, Joint Committee on Atomic Energy [Part 1] July 10-12, 17, 1968; Session 90-2 (1968). The complete statement regarding Article II violations is in Appendix B.

[30] Available at http://www.dni.gov/press_releases/20071203_release.pdf.

[31] Personal communication with Andreas Persbo, Senior Researcher, the Verification Research, Training and Information Centre.

[32] Those reports are available at http://www.iaea.org/Publications/Documents/Board/2003/gov2003-40.pdf and http://www.iaea.org/Publications/Documents/Board/2003/gov2003-63.pdf.

INDEX

H

I

J

K

L

M

N

O

T